DICTIONARY
OF
ELECTRICAL ENGINEERING

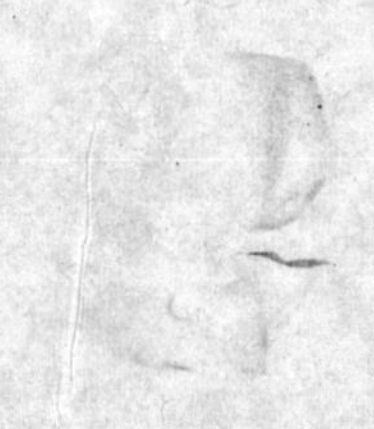

DICTIONARY OF ELECTRICAL ENGINEERING

K.L. Raina

ANMOL PUBLICATIONS PVT LTD
New Delhi-110 002

ANMOL PUBLICATIONS PVT. LTD.
Regd. Office: 4360/4, Ansari Road, Daryaganj,
New Delhi-110002 (India)
Tel.: 23278000, 23261597, 23286875, 23255577
Fax: 91-11-23280289
Email: anmolpub@gmail.com
Visit us at: www.anmolpublications.com

Branch Office: No. 1015, Ist Main Road, BSK IIIrd Stage
IIIrd Phase, IIIrd Block, Bangalore-560 085 (India)
Tel.: 080-41723429 • Fax: 080-26723604
Email: anmolpublicationsbangalore@gmail.com

Dictionary of Electrical Engineering

© Reserved

First Edition, 1989

Reprint, 1990, 1991

Second Edition, 1992

Reprint, 1993, 1994, 1995, 1996, 2010

PRINTED IN INDIA

Printed at Mehra Offset Press, Delhi.

Preface

This dictionary has been complied to cover terms which have been carefully selected from the various aspects of electrical engineering. This dictionary also includes terms from various fields which are allied to the electrical engineering industry. The selection of terms has been carried out mainly on the basis of current literature. Wherever necessary, line diagrams have also been added to make the terms more clear.

It has been difficult to decide what terms to omit. An effort has been made to keep the definitions concise, without omitting any necessary informations. This has been done with the help of extensive cross-referencing.

The dictionary will be of immense value to the diploma, degree and A.M.I.E. students.

When the dictionary of this kind is complied, it becames essential to refer the works of many experts in this field.

Lastly I must thank my publishers who have taken pains to publish this dictionary in a rather short span of time.

Comments and suggestions are most welcome.

K. L. RAINA B.E.

Absolute: Independent, unrelated. In temperature, for example, absolute zero is distinct from zero on an arbitrary scale.

Absolute Ampere: See ampere.

Absolute Permeability: Refers to the quotient of the magnetic flux density in a medium by the magnetic field strength. Absolute permeability is given by $\mu = \mu r \mu_0$, where μr denotes the relative permeability and μ_0 denotes the magnetic constant.

Absolute Permittivity: Refers to the quotient of the electric flux density in a medium by the electric field strength. Absolute permittivity $\varepsilon = \varepsilon r \varepsilon_0$ where μr denotes the relative permittivity and μ_0 denotes the electric constant.

Absolute Unit: Refers to the unit of any system that includes length, mass and time in its fundamental units. Systems of absolute electric units need an additional unit besides the basic mechanical units. This unit is representing either the magnetic to the electric property of space. The three systems of absolute electric units have been m.k.s., electromagnetic c.g.s. and electrostatic c.g.s.

Absorptiometer: See absorption meter.

Absorption:

1. The term used for conversion of radiant energy to a different form of energy when transmitted through a medium.
2. Refers to penetration of a fluid into a solid body.

Absorption Dynamometer: Refers to a type of testing brake which is used for measuring the output of electric motors and internal-combustion engines. It absorbs the output as well as measuring it. The simplest form has been the friction brake; the force needed to restrain the braking device and the distance of its line of action from the axis of the shaft have been measured, and the resulting restraining torque, equal to the output torque, has been calculated. Alternatively, a magnetic brake can be employed in which a metal disk rotates between the poles of electromagnets. For larger outputs the hydraulic dynamometer has been suitable and comprises a rotor, fitted with vanes, moving in an enclosed state or also fitted with vanes; admission of water to the device results in turbulence which provides a high braking torque.

Absorption Factor:

1. Refers to the ratio of radiation absorbed in a material to the radiation incident on it.
2. Refers to an allowance made in interior lighting calculations for light absorbed before reaching the working plane. In clear atmospheres the factor has been unity, but it has been less when smoke or steam is present.

Absorption Inductor: Depreciated term for current-sharing inductor.

Absorption Meter: Refers to an instrument which is used to measure the absorption of light transmitted through a sample of a transparent substance. It has been using a light-sensitive detector such as a photoelectric cell. An instrument of this type could be used, for example, for determining the concentration of a solution.

A.C.: Abbreviation of Alternating Current.

A.C.B.: Air-break (blast) Circuit Breaker.

A.C. Balancer: See static balancer.

A.C. Bridge : Refers to a bridge circuit which is using alternating current for the measurement of circuit parameters. Balance has been more complex in a.c. than in d.c. bridges, as both the magnitude and the phase angle of impedances enter into the balance condition. However, this fact has been able to make possible the precise measurement of reactive components (capacitors and inductors) as well as of resistors.

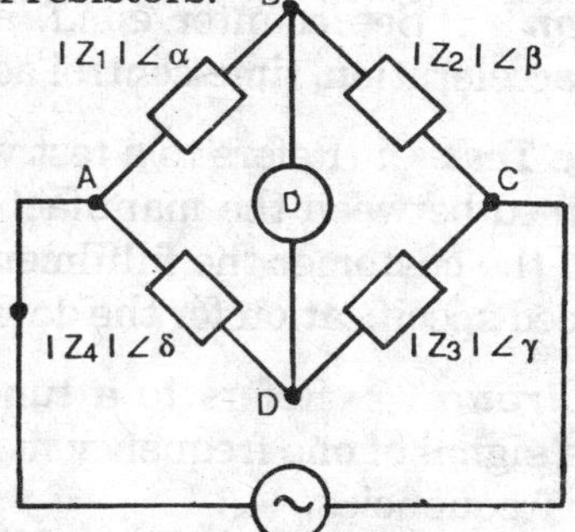

Fig.1 Four-arm a.c. bridge.

In the arrangement shown in Fig. 1 by analogy with the Wheat-stone bridge, balance would be obtained when

$$Z_1Z_3=Z_2Z_4$$

that is, when simultaneously

$$|Z_1Z_3| = |Z_2Z_4| \quad \text{and} \quad \alpha+\gamma=\beta+\delta$$

If $Z_1=R_1+jX_1$ be the unknown impedence, then

$$R_1+jX_1=(Z_2/Z_3)Z_4=(Z_2Z_4)Y_3$$

In the former relation, Z_1 gets balanced by a variable impedence Z_4 multiplied by a ratio (Z_2/Z_3); in the latter, it gets balanced by an admittance Y_3 multiplied by a product (Z_2Z_4). Examples of four-arm product bridges have been Maxwell bridge, Hay bridge and Schering bridge. Available for either type of measurement has been the universal bridge. Six-arm bridges have been the Anderson bridge and the parallel-T bridge (see also Heaviside–Campbell bridge).

A.C. Calculating Table : Refers to a form of network analyser.

Accelerating Machine : See particle accelerator.

Accelerating Relay : Refers to a device which is able to control the time interval between the closing of suc-cesssive resistor-short-circuiting contactors in starting a motor, so providing automatic acceleration. Such relays could be adjustable for time delay (timing relay), or of the current-operated type which closes when the current passing through it gets dropped to a predetermined value.

Acceleration : See counter e.m.f. acceleration, current-control acceleration, time-control acceleration.

Acceptance Test : Refers to a test which is contractually determined between the manufacturer and customer to prove to the customer the fulfilment of the conditions of the agreed specification for the device or equipment.

Acceptor Circuit : Refers to a tuned circuit which will accept a signal of one frequency more readily than those of other frequencies.

Acceptor Impurity : Refers to an element which is intro-duced into the meterial of a semiconductor, having a lower valency than the semiconductor. It captures, electrons to complete its valency bonding, producing a positive charge carrier or hole. This results in p-type conductivity.

Access Time : Refers to the time taken for extracting an item of information (a number or an instruction) from the store of a digital computer.

Accumulator : A device which is used for receiving and storing energy, and discharging it, by chemical action. Also called a storage cell or secondary cell, the ac-cumulator is reversible, *i.e.* it can, after discharging, be brought back to a full state of charge by passing a reverse current through it.

A storage cell is having essentially two electrodes (plates) which are immersed in an electrolyte in a suitable container. In practice, multiple plates may be used in the cell. There have been two basic types : lead - acid cell and steel - alkaline cell.

Acheson Furnace : Refers to an electric furnace of the non-melting type which is used for producing silicon car-bide. A heavy current is passed through a central core of coke, around which the coke and sand which make the silicon carbide get packed.

Acid Dip : An acid into which chemically cleaned materials have been dipped before electroplating in order to remove films of oxide, etc., which would interfere with the plating process.

A-class Insulation : Refers to one of seven classes of insulating materials for electric machinery and apparatus on the basis of thermal stability in service. Class A insulation has been assigned a temperature of 105°C. It consists of materials like cotton, silk or paper, suitably impregnated or coated.

A-connection : Refers to a three-phase transformer connection between phases which provides a neutral point.

Acoustic Controller : It is a form of automatic control by noise level. It is comprising a microphone, an amplifier, a high-pass filter to segregate the significant noise from low-frequency external noise, and a relay adjusted to operate in accordance with fluctuations in noise above and below a set level. An application has been the control of the feed motors in a pulverised fuel mill, by noise level inside the pulverising drums.

Acoustic Sounding : It is a system of determining the depth of the seabed or an immersed object, by time-interval measurements of the echo of acoustic waves, employing piezo-electric oscillators and recording instruments. The equipment is known as sonar.

A.C. Potentiometer : It is an instrument which is used for comparing alternating voltages. A.C. potentiometers are different from d.c. types as no alternating voltage standard comparable to a standard cell exists. In practice the a.c. potentiometer generally gets balanced on direct current against a standard cell, using a transfer instrument which has been commonly a precision electrodynamic galvanometer or milliammeter equally accurate on direct or alternating current. It is then used for adjusting the a.c. side for balance before use on a.c. measurements. Two general forms exist : the polar, and the quadrature or rectangular coordinate. An example of the polar type has been the Drysdale potentiometer, and of the quadrature type has been the Pedersen potentiometer.

A.C. Resistance : Alternative name for equivalent resistance, which has been preferred.

Acrylic Resin : Thermoplastic synthetic resin which is made from acrylic and methacrylic acids and used as insulating material. Polymethylmethacrylate (Perspex, Diakon, etc.) has been the most important; it has been a rigid, clear, transparent resin which can be obtained in flat or curved sheets, rods, tubes, and mouldings made from powder. It has been of value for high frequency use, but is having a low softening point which limits it to low-temperature applications.

Actino-therapy : Ultra-violet therapy, i.e. this term refers to treatment of disease by radiation with ultra-violet waves (0.40 – 0.25μm).

Active Circuit Element : Refers to a source of electric energy. An active circuit is having at least one active circuit element.

Active Component : Refers to the component of a sinusoidal current or voltage (considered as phasor quantities) which has been in phase with the voltage or current.

Active Electrode : Refers to an electrode of an electrostatic precipitator. The active (discharge) electrode gets insulated, and the other (receiving) electrode, usually positive and earthed, collects the precipitated particles.

Active Power : It is equal to the product of the current and active component of the voltage or of the voltage and active component of the current.

Actuator : It is an electromechanical device which is used for performing a mechanical action, generally with a linear rather than a rotary motion. The motion has been speed-and position-controlled in the same way as in a servomechanism.

Adaptor : It is a device which may get inserted into a socket-outlet to receive one or more pluges which may or

may not be of the same type as that for which the socket-outlet is intended. The name has been also sometimes applied to accessories for insertion into a lampholder.

Admittance : Refers to the quotient of current phasor by voltage phasor in a circuit with sinusoidal current and voltage. The SI unit of measurement has been the siemens (symbol: S).

Advancer : See phase advances.

A.F : Abbreviation of availability factor.

After-glow : Another name for persistence.

Ageing : Refers to the gradual change that takes place with time in the useful properties of a material.

Air-blast Circuit-breaker : It is a circuit-breaker in which the arc gets extinguished by a blast of air. It may be using compressed air to force the contacts apart as well as to blow out the resulting arc. Very high breaking capacities have been possible at alternating voltage from 6.6 kV to 380 kV. In all forms of air-blast circuit-breaker, high pressure air causes the arcing contacts to open, by pressure either on the contacts themselves, or on an actuating piston. A jet of air sweeped the arc into a low-pressure zone where it gets de-ionised and extinguished at a current zero. Arcing times of less than one cycle have been possible. Axial-blast, radial-blast and cross-blast jets have been employed, the last being used with arc chutes, for the lower voltages only. Arc chutes have been not needed for axial-and radial-blast circuit-breakers, but cooling vanes may be built into the exhaust chambers.

Air-break Circuit-breaker : It is a circuit breaker in which the circuit gets opened in air. It used 'free' air at atmospheric pressure to extinguish the arc and to provide insulation. Thus avoiding the need for processing of the insulant. Free-air circuit-breakers have been constructed for direct voltage up to 3 kV and for alternating voltages up to 3.3 kV in three distinct forms: low-breaking capacity *a.c* and *d.c.*circuit-breakers, high-speed *d.c.*circuit breakers and high-breaking cpacity *a.c* circuit-breakers.

Air Capacitor : Refers to a capacitor in which the insulant has been air.

Air Conditioning : Refers to the production of an artificial atmosphere which is suited to particular requirements. It may include cleaning, humidifying, dehumidifying, cooling or heating, air movement and possibly ozonising.

Air Cooling : The term used for the carrying away, by means of air, of the heat caused by inevitable losses in the operation of a motor, transformer or other machine. An open type machine may get cooled by natural convection, a protected type by fan or fans mounted on the shaft, and an enclosed type either by a fan mounted internally or externally, or by air blown through ingoing and outgoing ventilation pipes.

Air Core : Deprecated term. Refers to coreless.

Air Gap :

(1) Refers to part of the magnetic circuit of rotating electric machines and electromagnets, needing the largest proportion of the magnetomotive force; for example, the space between the rotor and stator of a motor, or between the armature and core of an electromagnet.

(2) Also, refers to a point of discontinuity in an overhead conductor for electric traction.

Air Termination : Refers to part of a lightning-conductor system which has been designed to collect discharges from the atmosphere or to distribute charge into the atmosphere.

Alarm Device : Refers to an instrument which is used to indicate by acoustic or visual means or both the existence of an abnormal situation in the operation of a device or apparatus or system.

Alcomax : Trade name for nickel-aluminium permanent-magnet materials which are exhibiting anisotropic characteristics. The properties in the preferred direction of magnetisation have been obtained by heat treatment in a magnetic field, and are very superior to the properties in other directions.

Alkaline Cell : See steel-alkaline cell.

Alkyd Resin : Synthetic resin, the condensation products of polybasic acids with polyhydric alcohols. They have been used (non-hardening resins) as adhesives, or (thermosetting types) as finishes or for bonding insulating materials such as asbestos or mica.

All in Tariff : Refers to an electricity supply tariff which does not involve any consideration of the purpose for which the supply is to be used. A domestic supply has been usually on a two-part tariff.

All-insulated : Having all external surfaces protected by coverings possessing entirely of insulating material.

All-or-nothing Relay : A monostable relay that gets changed to the energised condition when the power supplied exceeds an upper threshold value and changes to the non-energised condition when the power supplied drops below the lower threshold value.

All-pass Network : Refers to an a.c. network which is allowing power to pass at a low loss which is independent of frequency.

Allan Cell : American name for an electrolytic cell for the production of hydrogen by electrolysis of an alkaline solution.

Allen's Loop Test : It is a fault-localisation test which has been suitable for finding high-resistance faults in short lengths of cable. It has been similar to Varley's loop test.

Alni, Alnico : Trade names for nickel-aluminium permanent magnet materials which are exhibiting isotropic characteristic. Alni, was the first commercially used alloy, and has been now partially displaced. Alnico is having the highest magnetic energy of the isotropic alloys.

Alphanumeric Code : Refers to a code which is used in data processing where the characters include both letters and numerals.

Alternating Current, Voltage : A current or voltage which is varying with time in a cyclic manner, the current direction and the voltage polarity have been reversing periodically. Its main value has been zero.

Alternator : It is an *a.c.*generator having its field windings excited by means of direct current. Alternators could be divided into two types depending on the arrangement of the field system

(a) The turbo type where the surface of the rotor has been a smooth cylinder and the field winding has been kept in slots machined longitudinally in the steel.

(b) The salient-pole type having pole pieces which are surrounded by the field coils, projecting outwards from a hub.

Alumel : A nickel-base alloy having manganese, aluminium and silicon. It has been stable at temperatures up to 1200°C and has been used in thermocouples and as an electric resistance alloy.

Aluminium : It is a bluish-white metal having atomic number 13 and atomic weight 27. It is having properties of value in electric applications. Pure aluminium has been about one-third of the weight of iron, copper, zinc or their alloys. Its electric conductivity has been high, and its alloys have high tensile strength. It is corrosion-resisting.

Steel-cored aluminium conductors have been mainly used for medium and long-span lines and for very long individual spans. Aluminium wire, usually square, has been sometimes used for winding the coils of large magnets to save weight, and it can get insulated to withstand high temperatures. In the form of flat bar, rod or tube, aluminium has been especially used for busbars. Other applications of pure aluminium have been for electricity meter disks, and, as pressed or drawn sheet, for electrostatic screening.

Aluminium Alloy : It is an alloy of aluminium combined with another metal or metals, principally copper, magnesium, zinc, silicon, manganese or nickel, to improve its characteristics. Alloy with 5-12% silicon has been used for busbar casings, since it has been non-magnetic, light, and has good casting and reasonable machining properties. High-tensile alloys such as Duralumin have been employed for the moving parts of switch-gear, and result in a reduction of inertia effects. Silicon alloys used for castings for cage rotors, the percentage of silicon varying between 5% and 12% depending on the resistivity required. Aluminium alloys having 2.5-5% copper find use for the collector bows used with overhead wires by electric trolleys and locomotives.

Aluminium Rectifier : It is a type of electrolytic rectifier which is having cells each with an electrolyte of ammonium phosphate, having an aluminium anode and an inert lead-plate cathode. Such a cell has been a unidirectional conductor. The action has been chemical as it depends on the forming and reforming of a film on the anode surface. There occurs appreciable reverse current and the efficiency has been a low. Only low voltage and currents can get handled by a single cell, but a number may be connected in series or parallel.

Ammeter : An ampere-meter, which is used for measuring current, having a scale and moving element which carries a pointer. It get connected in series with the circuit carrying the current to be measured and has been of low electric resistance or impedance, in order that only a small voltage drop shall get caused in the circuit and a minimum of power absorbed from it. Ammeters have been moving-iron, moving-coil, electrodynamic, hot-wire, induction or rectifier instruments.

Ammeter Shunt : Refers to one or more flat strips of metal alloy, of very low temperature coefficient *e.g.*manganin), and of definite resistance, connected in the main circuit and across the terminals of a moving-coil millivolmeter which has been calibrated as an ammeter.

Ammonium Dihydrogen Phosphate : It is a material which is generally known as ADP. It shows pronounced piezo-electric effects. It is used as a transducer element for ultrasonic depth measurement and in submarine detection systems.

Amortisseur : Refers to the damping winding of a machine.

Ampere : The SI unit for electric current (symbol A)

Ampere Balance : It is a laboratrory currrent=measuring device. A sliding weight balances the force between a pair of coils carried on the weight beam and a system of fixed coils through which the same current have been passing.

Ampere-hour : It is a unit of electric charge (symbol Ah). It has been the quantity of electric charges that passes in one hour at a current of one ampere, 1 Ah=3600 coulombs.

Ampere-hour Capacity : Refers to the capacity of a battery which is expressed in ampere-hours usually based on continuous discharge at a speciefid rate.

Ampere-hour Efficiency : Refers to the ratio between the output quantity in ampere-hours from a battery during a test discharge input Ah requried to charge the battery.

Ampere-hour Meter : Refers to a direct-current electricity-supply meter which is able to measure the current-time summation, or ampere-hours.

Ampere-turn : It is the SI unit for magnetomotive force (symbol: At).

Ampere's Law : See Maxwell's laws (I).

Amplidyne : Trade name for a range of rotating amplifiers of the cross-field excited type. It has been quick-response d.c. generator, the output of which has been controlled by a very small field power. Because of this property, the Amplidyne has been particularly suitable for use as an exciter in a closed loop control system.

The method of excitation could be understood by reference to Figure 2 If current has been supplied to the control-field winding, a flux (indicated by two dotted arrows) would be set up on the vertical axis and induces an electromotive force in the armature conductors between the cross brushes. As these brushes have been joined together the e.m.f. makes a current to flow, and the passage of the quadrature current through the armature windings would set up an armature-reaction field (indicated by the two solid arrows). This field has been used to induce the main output voltage of the machine between the other pair of brushes shown on the vertical centre-line.

The e.m.f. needed to pass the quadrature current through the armature winding and the short-circuited

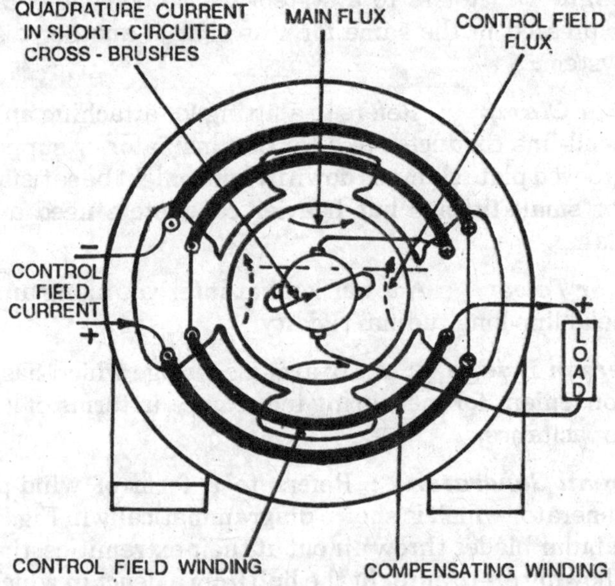

Fig.2 Internal connections of amplidyne

corss brushes has been small compared with the full output voltage of the machine, and the power needed by the control field winding has been therefore only a small fraction of that which would be needed to excite the machine in the normal way. In a fully compensated machine the power amplification ratio could be very large (20,000 or more).

Amplifier : It is an apparatus which is used for producing a magnified version of an input signal. The term has been commonly applied to an electronic circuit which is used for amplifying a voltage applied to its input terminals.

Amplitude : Refers to the peak value of a sinusoidal quantity.

Amplitude Distortion : Refers to a change in waveform, when deviations in the ratio of the r.m.s. value of the output to that of the input occur for different input amplitudes due to a non-linear response in the system.

Amplitude Modulation : Refers to the imposition of a signal shape on to an alternating carrier wave by variation of the carrier amplitude. The maximum signal frequency be appreciably less than that of the carrier.

Analogue : Refers to a system whose behaviour can be expressed in the same form as that of another distinct system.

Anchor Clamp : Refers to a fitting for attaching an overhead-line conductor to a tension insulator or support. A grooved plate clamped down by bolts might be satisfactory for small fittings but heavier conductors need a snail clamp.

Anchor Tower : A tower kept at intervals along an overhead line longitudinal rigidity.

Anderson Bridge : A six-arm a.c. bridge which has been convenient for measuring inductance in terms of a fixed capacitance.

Andreau Generator : Refers to a form of wind-power generator which is shown diagrammatically in Fig.3. The tabular blades throw air out at their extremities, thereby allowing air to enter at the hub from a deuct in which has been placed an air turbine driving a generator. This scheme has been theoretically less efficiency than the more common propeller type, but is having the advantage that the generating equipment gets fixed and located at ground level.

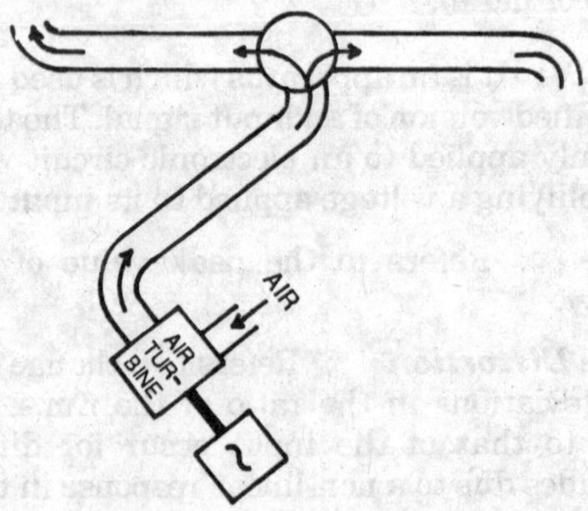

Fig. 3. **Andreau wind-power generator. As the wind drives the blades round, air is thrown out icentrifugally from the blade tips.**

Angstrom : It is a unit of length which is equal to 10^{-10}m or 10^{-4}μm. It is used in measuring the wavelengths of short electromagnetic waves, especially in the X-ray spectrum (symbol : A).

Angular Frequency : Refers to the measurenent of frequency of a regular recurring phenomenon. It may be expressed in radians per second, i.e. $w=2\pi f$ rad/s, for a frequency f in hertz. It is corresponding to the angular velocity of rotation in radians per second of a two-pole synchronous machine, and to the time angle in radians per second of a time-varying sine wave of frequency f.

Anion : It is a negatively charged ion, which exists in an electrolyte or a gas conduction. It will travel towards the positive electrode or anode under the influence of an applied voltage.

Anisotropic Conductivity : Refers to property possessed by a body which is having different conductivity for various directions of current flow.

Anisotropic Magnetism : Refers to the property which is exhibited by certain alloys when heat-treated in a strong magnetic field. They develop extremely strong magnetic properties in the directions of the field at the expense of those in other directions. This gives rise to very powerful and stable magnets.

Anode : Refers to the electrode out of which an electric current flows, with the convention that the current directions may be defined in terms of the movement of positive charges. The term has been applied typically to the appropriate electrodes of electron tubes, murcury-arc rectifiers and electrolytic baths.

Anode Drop : In an electric welding machine this term refers to the differences in voltage between the positive end of the arc-stream and the adjoining conductror.

Anodic :

(1) Possessing a more positive electrode potential.

(2) Indicating an element above hydrogen in the electrochemical scale.

Anodic Etching : Refers to a preparatory process which is used for deposition by electroplating. The metal to be plated has been made the anode in a suitable electrolyte.

Anodising : The production, on the surface of parts made of aluminium or its alloys, of a protective film of hydroxide. The part has been made the anode of an electrolytic bath, and a current is passed.

Anolyte : Refers to the portion of an electrolyte which is surrounding the anode, which is affected chemically by reactions taking place at the anode.

Antiferromagnetism : Refers to phenomenon which is exhibited by materials in which, in the absence of an applied magnetic field, the magnetic moments of adjacent atoms or ions have been in opposition such that the resultant magnetic moment has been zero but align with increasing field. Their susceptibility passes through a maximum as the temperature increases.

Antinode : Refers to the appearance of maximum value of a standing wave.

Aperiodic : Applied to an oscillatory system where excessive damping does not allow free oscillation.

Apparent Power : In an a.c. circuit this term refers to the product of the r.m.s. values of voltage and associated current.

Apparent Resistance : Obsolete alternative term for impedance.

Arc-back : Deprecated alternative term for backfire

Arc Chute : Device which is used in an air-break circuit-breaker to confine the arc and prevent it from spreading to adjacent metal work. A typical form has been shown in Figure 5. Several slotted plates of insulating material as shown in:

(a) Get arranged between two side plates so that a sectional plan on the plane PQ forms the labyrinth.

(b) The contacts have been located in the lower portion of the slot so that, after striking, the arc rises up in the slots, eventually taking up the zig-zag form shown in (b) and becoming lengthened and cooled as required.

Arc : It is a luminous column of ionised gas which is forming gas conduction. Any gas or vapour will become an electric conductor if it gets heated to a sufficiently high temperature, and when cooled again will become quite a good insulant. An arc, shown in fig. 4 has been generally divided into three parts in series; the cathode, the column or plasma, and the anode.

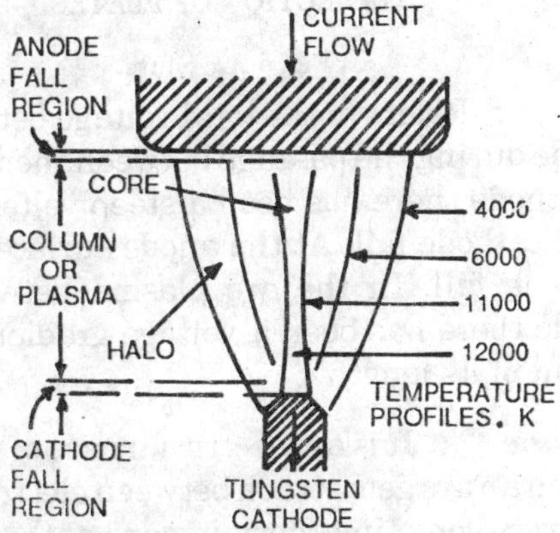

Fig.4. A characteristic electric arc.
The arc shown is of 200 A in nitrogen
in atmospheric pressure.

Arc-control Device : Refers to are equipment which is used for controlling the arc which inevitably occurs between the contacts of the interrupting device when a circuit is interrupted. Almost all circuit-breakers must get fitted with are-control devices designed to keep the duration of the arc to a value between about 10 and 100 ms.

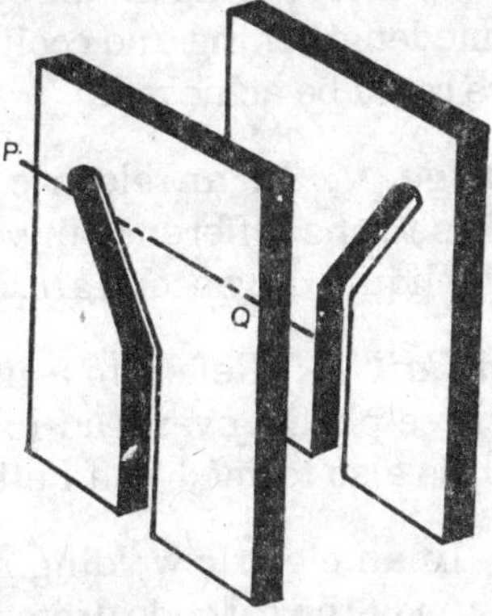

(*a*) SPLITTER PLATES

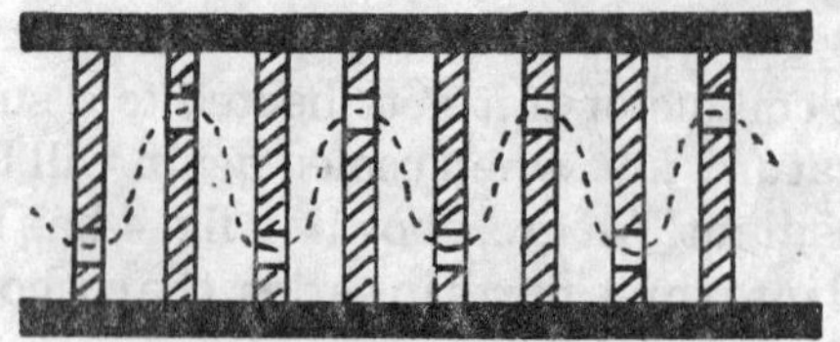

(*b*) SECTION OF PLATES

Fig. 5 Arc chute.

Arc Drop : Refers to the total voltage between anode and cathode during the passage between them of an arc. Near the cathode there has been a steep voltage gradient and a total cathode fall At the anode surface there has been an anode fall. In the arc plasma between anode and cathode there has been a voltage gradient roughly independent of its length.

Arc Furnace : It is an electric furnace of the melting type in which an arc gets struck between electrodes, the metallic charges sometimes forming one of the electrodes.

Arc Rectifier : Refers to a rectifier having a cathode and an anode between which an arc is maintained.

Arc Splitter : Refers to an arc-control device comprising an arc chute in which the side plates get separated by metal plates or pins. As the arc travels up it is split into a number of smaller arcs in series. At the anode and cathode of any arc there has been significant voltage drops, so that by introducing a number of such drops, as well as by some lengthening and cooling, a high effective arc-resistance could be achieved.

Arc-stream Voltage : In an electric welding machine, this term refers to the difference in voltage between the two ends of the liquid or gaseous arc.

Arc-suppression Coil : Refers to a neutral-earthing inductor in a three-phase, overhead-line, power-distribution system. It is also termed as a Petersen coil.

Arc Voltage : In an electric welding machine, this term refers to the sum of the cathode drop, arc-stream voltage and anode drop.

Arc Welding : Refers to fusion welding, the heat necessary to melt both the parent plate and filler material being derived from an electric arc. The process may be involving either a consumable or a non-consumable electrode. In consumable electrode processes, the arc has been struck between the parent plate and an electrode which melts off to form the filler metal required in the joint. In non-consumable electrode processes, the arc has been struck between a non-consumable electrode and the work-piece, with filler metal fed separately into the joint if needed.

Arcing Contacts : Refer to contacts which are incorporated in switchgear to protect the main contacts from any deleterious effects due to arcing. They open after and close before the main contacts. They have been made arc-resisting by suitable choice of material (e.g. carbon). They are also easily renewable.

Arcing Horn : Metal projections which are placed at ends of overhead-line insulator strings to provide an alternative path for an arc which might otherwise damage the insulator units.

Arcing Shield : See grading shield.

Argon Arc Welding : Refers to the commonest form of non-consumable electrode arc welding. It employs a tungsten electrode. A shield of inert gas, usually argon but occasionally helium, protects both the weld pool and the electrode from atmospheric contamination.

Armature :

(1) Of a machine, armature refers to one of the main members in which rotational electromotive forces are induced, and which has been associated with the conversion of electric to or from mechanical energy, in a.d.c. machine and an induction machine the armature has been almost invariably the rotating member; in a synchronous machine it has been normally the stationary member (or stator).

(2) Of an elecromagnetic circuit, a movable part which, by attraction, can be made to do useful mechanical work, as in a relay or contactor.

(3) Of a permanent-magnet.

Armature Head : The term used col for the end-plate of a laminated armature core.

Armature Reaction : Refers to the effect on the total working flux in an electric machine of the armature magnetomotive force which is produced by armature currents.

Armouring : Refers to the steel wires or tapes which are wrapped round a cable for mechanical protection.

Aroclor : A chlorinated diphenyl which is used as a dielectric in industrial capacitors.

Astatic System : Refers to an arrangement of magnets or coils which are so designed that no resultant force has been exerted on it by an external uniform magnetic field.

Asymmetrical Breaking Capacity : Refers to the r.m.s. value of the combined a.c. and d.c. components of current that can be broken at a stated voltage by any one pole of a circuit-breaker.

Asynchronous Compensator : Refers to an asynchronous machine which is used for power-factor correction.

Asynchronous Machine : Refers to an a.c. electric machine whose normal operating speed (unlike that of a synchronous machine) has not been restricted to the synchronous speed. The operating characteristics have been however, functions of the speed-difference, or slip, by which the operating speed get departed from the synchronous.

Atmoshperic Electric Charges : Atmospheric charges which are existing in the Earth's atmosphere. The voltage gradient near the ground has been of the order of 150 V/m in fine weather, 1.5 kV/m in fog, 10 kV/m in thunderstorm conditions.

Atomic Hydrogen Welding : Refers to a form of non-consumable electrode are welding which uses two tungsten electrodes between which the arc is struck. A stream of hydrgoen gets directed across the arc and the molecules of hydrogen break down into atoms. As the gas stream passes out of the arc the atoms of hydrogen recombine and the exothermic reaction produces a very high temperature 'flame' with which the welding has been carried out. The hydrogen burns as it leaves the arc, thereby protecting the weld zone from atmospheric contamination. It is also termed as atomic arc welding.

Attenuation : Refers to a reduction in the magnitude of a quantity; especially that reduction in the amplitude of a voltage or current due to its transmission over a line or through an attenuator. In each case the series resistance and shunt conductance losses get reduced the signal energy during its passage.

Attenuation Factor : Refers to the natural logarithm of the ratio of input to output r.m.s. voltage or current in a transmission line or an attenuator when get terminated with a matched load. Thus with V_1,I_1 and V_2,I_2 as input and output voltage and currents.

$$V_1/V_2=I_1/I_2=\exp \alpha$$

where α denotes the attention factor with the unit neper.

Attenuator : Refers to a device which is used for introducing a specified loss between a signal source and a matched load without upsetting the impedance relationship necessary for matching. Resistors arranged in T or II formation, called attenuator pads, may be used for this purpose. Fig. 6 Exhibits at: (a) A simple symmetrical-T pad. If the attenuation to be adjustable in steps, the bridged-T form shown in (b) has been preferred, the variation being obtained by switching R_1 and R_2 simultaneously. More

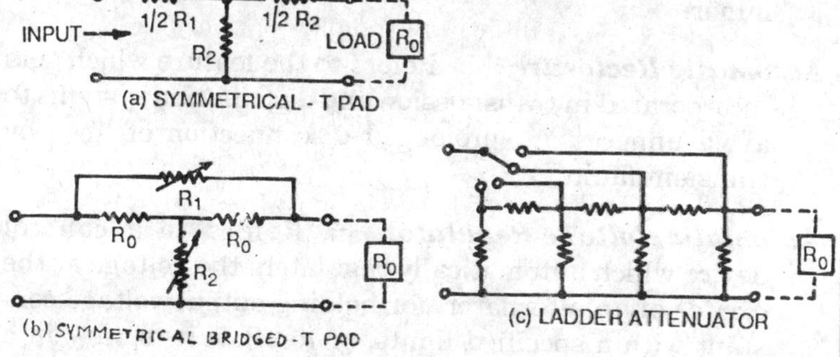

Fig.6. Attenuators.

economical from the view point of simple switching has been the ladder attenuator (c) but on the minimum-attenuation position (at the right-hand end) the network still shunts the load, providing a loss of 3 dB.

Attracted-armature Relay : A relay having an electromagnet, the field of which gets influenced an armature arranged to operate the contacts. When the coil gets energised the force on the armature would be proportional to (ampere-turns/gap-length)^2 and hence increases as the armature moves to reduce the gap.

Audiometer : An instrument which is used for diagnosis of hearing ability and otological research. It has been an accurately calibrated variable-frequency audio oscillator producing pure sounds over the range 50 Hz - 20 kHz. The amplitude of the signal can get varied, and in testing auditory acuity the patient has been asked to indicate when the sound appears and disappears for various frequencies.

Automated Equipment : Equipment used in mass production process which are designed to carry out a fixed routine.

Automatic Control : Refers to the control of a device, system or manufacturing process with the minimum of human intervention. It has been usual to exclude from this category the many forms of semi-automatic operation (as, for example, automatic passenger-operated lifts) that need some form of human initiation, even if it be only minor.

Automatic Reclosure : Refers to the feature which gets incorporated in transmission and distribution systems to avoid unnecessary prolonged disconnection of lines by transient faults.

Automatic Voltage Regulator : Refers to a control device which automatically regulaters the voltage at the exciter of an alternator, for holding output voltage constant within specified limits, or to allow it to vary in a predetermined manner. There have been four main types: rheostatic, pulse, normally inactive and static.

Automation : Refers to the automatic control of manufacturing and other process through all their stages, which involves self- examination and feedback of corrections.

Auto-Transformer : A type of transformer which is having only one winding per phase, part of this winding being common to both primary and secondary sides. This makes an appreciable saving in first cost and higher operating efficiency compared with a two-winding transformer, provided that the the ratio of transformation has been not greater than about 2. A single-phase auto-tranformer is consisting of a single winding. The most widely used three phase autotransformer has been star-connected, as shown in Fig. 6, 7. A neutral point has been available for giving a four-wire supply. The autotransformer is having a number of disadvantages

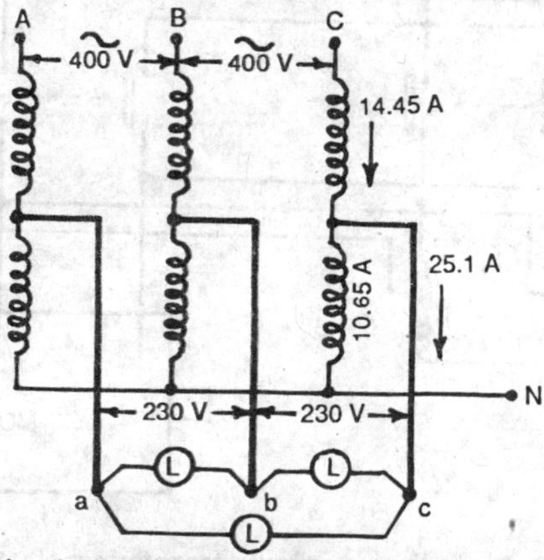

Fig.7 . Connection for 10 KVA 400/230 V three-phase auto-transformer.

which restrict its use. As there has been only a single winding, its inherent reactance has been low, and it has therefore restricted self-protection against damage from short-circuits. Due to the electric continuity of the windings the higher voltage has been liable in the event of a fault to become impressed on the low-voltage system, which makes added insulation and additional precautions necessary.

Auto-transformers are used in voltage-regulating devices, for processes involving a variable control, like the loading on furnace transformers, and for supplying a reduced voltage when starting up induction and synchronous motors.

Auto-transformer Starter : Refers to a tapped auto-transformer by means of which reduced voltage could be applied to a motor to limit the current taken during starting. The motor could be started by connecting its primary to tappings on the starting transformer. After a time delay, it has been re-connected direct to the supply. The basic diagram has been shown in Fig.8. The simple auto-transformer starter is having the dis-advantage that at the instant of transition from start to run, the supply to the motor is interrupted. It implies that the insulation may be stressed by high transient voltages. It has been avoided in Korndorfer starting.

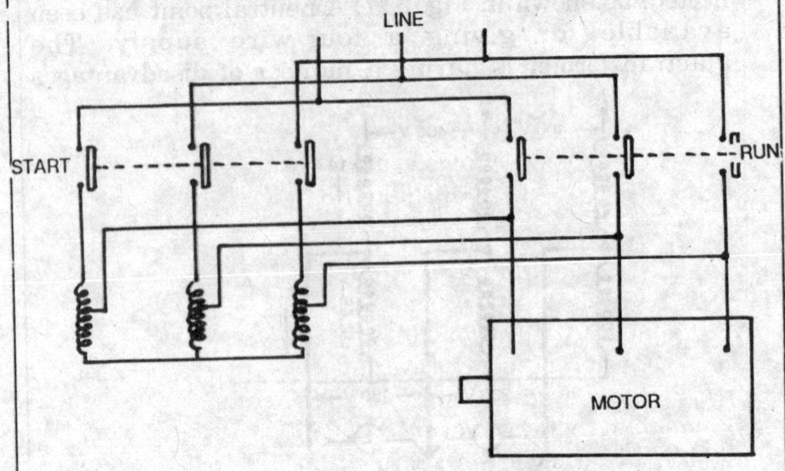

Fig. 8 . Basic diagram of auto-transformer starter.

Auxiliary Contacts : Refers to contacts in a contactor which are used for operating auxiliary devices or for carrying the circuit current during make and break.

Auxiliary Plant : It is a plant which is supplied to a generating station, such as feed pumps, oil pumps, circulating-water pumps, coal and ashhandling gear, switch tripping and lighting etc. the requirements depending on the type of station.

Availability Factor : In a power system or part there of, availability factor refers to the ratio of (a) the duration during which a particular item or section is available for normal operation, to (b) the whole of a specified period.

Avalanche Breakdown : In a. p. n. junction this term refers to the electric breakdown which is caused by cumulative multiplication of free charge carriers under the action of an excessively strong electric field.

Avalanche Transistor : Refers to a junction transistor in which the collector has been operated into the current multiplication breakdown region. This provides a current gain greater than unity and producers a negative resistance region.

Average Value : See mean value, rectified value.

Ayrton Shunt : It is a form of Shunt which is designed for open-circuit work or for use in circuit high resistance.

B-Battery : It is a high-tension battery. It is having one or more blocks of dry cells. It is used for the anode circuit of a thermionic tube.

B-class Insulation : Refers to one of seven classes of insulation materials for electric machinery and apparatus on the basis of thermal stability in service. Class B insulation has been assigned a temperature of 130°C. It consists of material like mica, glass fibre or asbestos suitably impregnated or coated.

B/H Curve : Refers to the graph of the relation between the magnetic flux density B and the magnetic field strength H. The quotient B by H gives the absolute permeability, which for a ferromagnetic material has been a function of H. Fig.1 shows for a permanent magnet material a closed B/H curve in the form of a normal hysteresis loop. If the curve gets extended into magnetic saturation as demonstrated in Fig.1 then B=OA would be the magnetic remanence, H=OC would be the coercivity and ADC would be the demagnetisation curve.

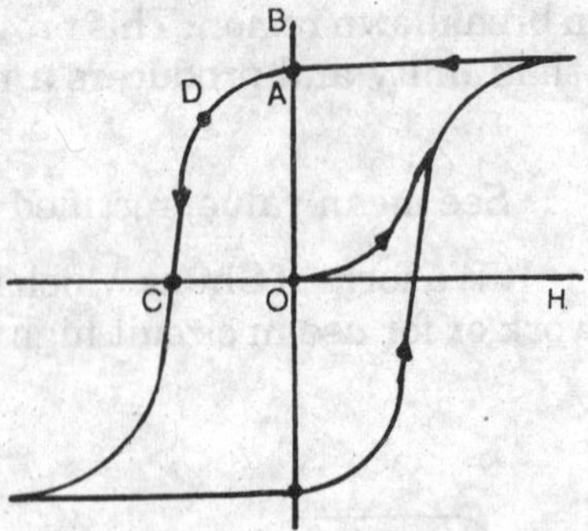

Fig. 1. Hysteresis loop of a permanent-magnet material.

Back e.m.f. : The term used by the voltage set up in a circuit which is opposing the applied voltage. For example, a back e.m.f. is produced in the armature conductors of a d.c. motor due to the rotation of the armature in its magnetic field.

Back-pressure Plant : Refers to the steam-driven generating sets which are delivering exhaust steam at a positive pressure suitable for industrial process work.

Back-to-back Test : Refers to a method of testing electric machines or transformers under full-load conditions having only a small consumption of power. Two identical units have been connected so that one unit supplies power to the other from which the output gets returned to the input of the first unit. It has been only necessary to provide a means to supply power from an external source to make up for the losses in both units.

There are three methods of making back-to-back tests on series motors which are shown in Fig. 2 In method (a) a variable voltage booster has been added to the generator circuit to raise its voltage slightly above the motor voltage, so that the output of the generator can be returned to the supply. Method (b) has been using an auxiliary driving

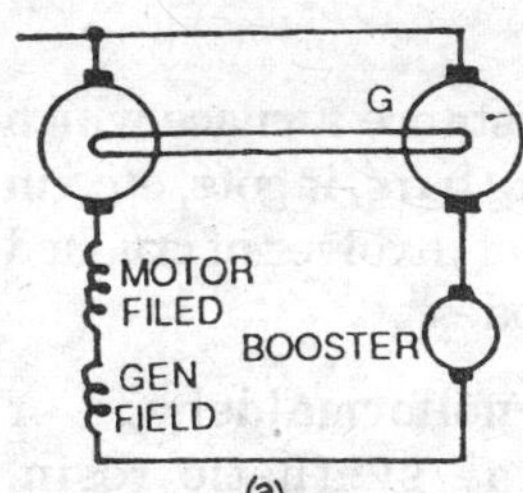

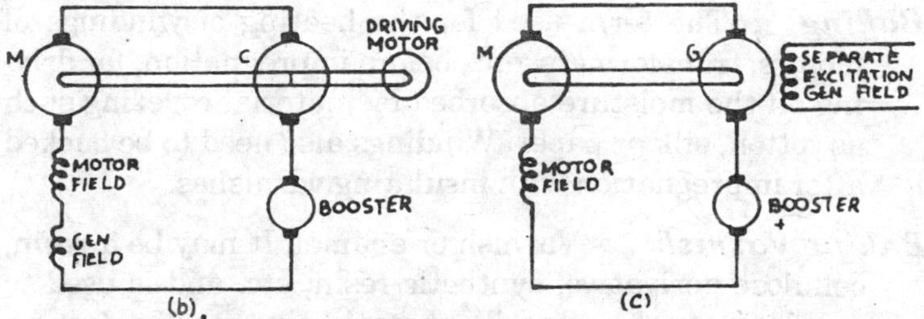

Fig. 2. Back-to-back tests on series motors.
The motor acting as a generator (G) must
be similar to the supplied motor (M)

motor, which supplies the iron, friction and windage losses, and a low-voltage booster which supplies the copper losses. Method (c) has been using a separate source of excitation, although on small motors the field can be put in series with the generator and controlled by a diverter.

Back-up Protection : The term used for a scheme of equipment protection, the stability of which has been dependent on the operation of protective system of other apparatus, and the operation of which extends over a range of protected or unprotected apparatus.

Backfire : In the operation of a mercury-arc valve this term refers to the ignition of an anode when at a negative voltage with respect to the cathode. This brings about a short-circuit current from the a.c. power supply through the valve, and thus a fault in operation.

Backward Lead, Backward Shift : Of the brushes on a motor commutator, this term refers to their angular displacement from the neutral open-circuit position in a direction opposite to that of the rotation of the commutator.

Becon Cell : It is a fuel cell which is using hydrogen and oxygen as the reactants.

Bailey Furnace : It is a type of resistance furnace which is used for annealing and for heating bars, ingots, etc., in rolling mills. The resistance material involves of crushed coke packed between carbon electrodes.

Bakelite : Trade name for a phenolformaldehyde or cresolformaldehyde thermosetting synthetic resin. Bakelite has high thermal and electric resistance.

Baking : The term used for the heating of windings of motors, transformers, etc., before impregnation, for driving off the moisture absorbed by material covering such as cotton, silk or paper. Windings also need to be backed after impregnation with insulating varnishes.

Baking Varnish : Varnish or enamel. It may be a resin, cellulose derivative, synthetic resin, etc. and is used to provide a tough and resilient coating in the manufacture of insulating materials, or for the protective and insulating impregnation of windings. The evaporation of solvents and the hardening of the materials has been carried out by baking at, say 100°C for several hours.

Balance : This term finds use in connection with bridge measurements. A balance is obtained when the impedances forming the arms of the bridge get adjusted so that no current flows through the galvanometer.

Balanced Current Protection : Refers to a form of system protection arranged for defecting phase faults and earth faults or, more simply, earth faults only. In both cases current transformers get mounted on each circuit connected to the section being protected. These are having their secondary windings connected in parallel through wires across which a relay is connected. Under healthy conditions the balance of primary current has been reproduced in the secondary circuit and the relay has been unenergised. During internal faults, the current-balance condition fails and the relay operates. By suitable switching of the secondary circuits complete discrimination between faults on the various sections in achieved for all normal operating conditions.

Balanced Load : Refers for the symmetrical arrangement of loads on a system. A d.c. three-wire system has been said to be balanced when the load has been equally shared between the middle wire and the outers. An A.C. polyphase gets balanced when the loads taken from each phase have been equal and at the same power factor.

Balancer : Refers to a device which is used to maintain, automatically or otherwise, a balanced load in an a.c. or d.c. system. It may also be used for converting a single-phase supply from two or three-wire, or a three-phase supply from three to four-wire.

Balancing Battery : Refers to a floating battery which helps a generator for supplying, current on peak loads. It gets connected in parallel with the generator busbars, discharging during heavy load and charging during light load. At normal load, the current has been supplied entirely by the generator, the battery being neither charged nor discharged.

Balancing Machine : The term used for a machine which is used for determining and measuring by electric means the amount and position of the unbalance forces in bodies like armatures and rotors while rotating.

Balancing Ring : Ring conductors attached to an armature, to which equipotential or equalising connections have been made, to allow an interchange of current so that currents in different parts of a lap armature winding have been kept symmetrical.

Ballast : Refers to a device, such as a resistor or inductor which gets inserted in series in a circuit to limit the current and provide stable operation under changing conditions.

Ballistic Galvanometer : It is a galvanometer having a moving system which has a considerable moment of inertia and a low damping factor, so that the time of swing has been long compared to the duration of the current pulse it measures. The swing has been then proportional to the current-time integral of the pulse.

Bank : Refers to several similar electric devices so connected that the combination functions as an integral unit.

Bar Suspension : Refers to a method of mounting a traction motor on an electric truck. One side of the motor gets supported on the axle by using suspension bearings and the other by lugs projecting from the frame and bolted to a spring-suspended bar lying transversely across the vehicle. This is also formed as yoke suspension.

Bar Winding : Refers to an armature winding which is consisting bars of rectangular section. The winding has been made of straight copper strips, insulated on the slot portion, forming a half-coil. One end has been bent to shape and the half-coil gets pushed through the slot. It has been then bent at the other end and joined at both ends to form the winding.

Barium Titanate : A meterial having piezo-electric properties. A polycrystalline ceramic of barium titanate and various binding materials acts as a single crystal by the application of an electric field, the direction of the field deciding the direction of piezo-electric activity.

Barkhausen Effect : Refers to an effect which is observed when a ferromagnetic material is subjected to a change in an applied magnetic field. When the magnetising force gets increased, small discontinuities appear in the magnetisation curve of the substance, because of reorientation of the domains.

Barrel Plating : The term used for an electroplating process in which the articles to be plated are kept in a rotating barrel or container.

Barrel Winding : An armature winding in which the end connections lie flat upon a cylindrical surface.

Barretter : Refers to a resistive element which is having a high positive temperature coefficient of resistance. It is used to maintain the current in circuit approximately constant in spite of variations of voltage. The barretter operates by varying the resistance of the circuit proportionately with the voltage. Iron and nickel are suitable materials and one of these may be employed as a filament in an atmosphere of hydrogen or helium.

Base Load : Refers to the minimum demand on a power supply system. It has been met as far as possible by the more efficient generating stations, the rest of a peak load being supplied intermittently by the less efficient stations.

Base Plate : See bed-plate.

Basket Winding : The term used for a distributed winding for a.c. machines. The name has been derived from the appearance of the linked coils. It is also called a chain winding.

Batch Furnace : See resistance furnace.

Battery : Refers to an assembly of similar units; in particular a series connection of voltaic primary cells or secondary cells. Batteries of secondary cells are also termed as storage batteries or accumulators.

Battery Vehicle : Refers to a road or rail vehicle which gets self-propelled by means of electric motors supplied from an accumulator.

Bayonet Cap : It is a lamp cap in which the cylindrical outer wall carries pins (usually) two or three in number) for engaging in slots in the lampholder. It normally has two end contacts insulated from the outer wall. These caps may be referred to as BS (bayonet cap) or SBC (small bayonect cap) depending on size.

Bearing Current : Refers to a stray current which flows between the shaft and bearings of an electric machine.

Beat-frequency : Of two sinusoidal oscillations having a relatively small frequency difference, the difference of the two frequencies when the oscillations get combined.

Bed-plate : Base on which machine frame and bearing have been mounted.

Bedding : Of an armoured cable, a fibrous layer on which the armouring has been applied. The bedding may be permeated with a waterproof compound.

Bel. : Refers to the common logarithm of a power ratio. The numeric has been very rarely used; but in the form 10N, *i.e.* in decibel units, it has been commonly employed in communication technology.

Bell : Refers to a sounding apparatus in which a hammer for striking a bell or gong has been operated by an electromagnetic mechanims. This mechanism may be arranged to give either a single-stroke or a trembling action.

Bell Transformer : Refers to a two-winding transformer which may get connected to the mains supply to supply a bell circuit. A choice of outputs, 3, 5 or 8 V. or in larger sizes 12 or 24 V, has been generally provided from the terminals on the secondary side. The transformer has been permanently connected to the mains, as the magnetising current taken when idle have been very small.

Belted Cable : Refers to a form of power cable in which three insulated cores have been laid up together (with wormings to fill the interstices), and the whole assembly

Fig. 3 Belted cable.

has been then wrapped with a 'belt' of paper tapes (see Fig. 3). The design provides excellent service up to 11 kV, but above this voltage breakdown can occur due to tangential electric stresses.

Bentley - Galloway Discriminator : Refers to a crane load discriminator which is having an electric motor driving a centrifugal brake connected to the load motor.

Betatron : Refers to an orbital accelerator in which electrons have been both directed into circular orbits and also accelerated by using a magnetic field.

Bifilar Suspension : Refers to suspension of the moving part of an instrument by two vertical wires or threads which provide a considerable controlling torque. The period of torsional, T has been given by $T=4\pi(11/mgd^2)$ where I denotes the moment of inertia of a suspended body of mass m, and the threads have been of length 1 and distance d apart.

Bifilar Winding : Refers to a winding which is comprising two insulated conductors which are arranged side by side and connected in series so that a current through them passes in opposite directions. The inductance of such a winding has been generally negligible.

Bifurcating Box : Refers to a closed box in which the two cores of a twin cable can get connected to external conductors.

Bimetal : A temperature sensitive device which is produced by firmly bonding together two or more metals or alloys having sensibly different coefficients of expansion. The composite materials gets rolled into a flat strip, from which a variety of shapes may be cut. A change of temperature causes a change of curvature, utilisable in several forms such as the deflection of a straight strip, the opening or closing of U-shaped pieces, the rotation of spirals and helices, or the 'dishing' of washers and disks.

Bimetallic Instrument : Refers to an indicating instrument involving a bimetallic strip which gets fixed at one end while the other end operates a pointer. Current to be measured may be passed through the strip, but usually there has been a separate heater, with or without a thermal barrier to control the heating rate. The instrument has been sometimes used as a maximum-demand ammeter.

Biot-Savart Law : This law may be stated that the magnetic field strength H, at a point P (Fig. 4) due to a current I flowing in a network, has been the same as if each element dl of the network contributed a vector

$$dH = I\,dl\,\sin\theta / 4\pi s^2$$

The direction of this contributory vector has been at right angles to dl and to s, in such a sense that dH forms-handed screw about the direction of the current.

Bipolar Electrode : In an electrolytic cell, this is an additional electrode, which gets insulated externally from the anode and cathode, which is in the path of the current. Various parts therefore act as anodes, and parts as cathodes.

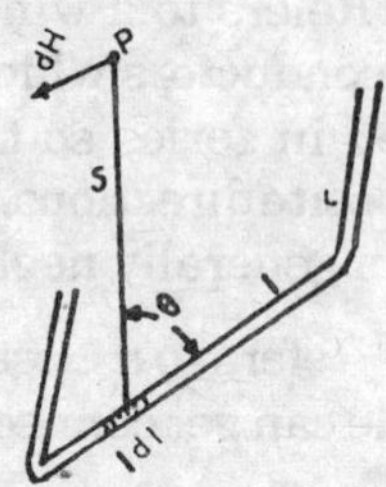

Fig. 4. The magnetising force due to the current in a conductor.

Bipolar Machine : Refers to a machine in which the field magnet is having only two poles. It is also called a two-pole machine.

Bismuth Spiral : It is a flat spiral of bismuth wire which is used for measuring the intensity of a magnetic field. The resistivity of bismuth gets increased in the presence of a magnetic fields, and so the change in resistance of the spiral can be employed for measuring field intensity (*e.g.* in the short air-gaps of electric machines) after calibration in known fields.

Bistable Relay : Refers to a relay having two non-energised rest conditions. When responding to an energing quantity it gets changed from one rest condition to the other where it remains when energisation gets removed. Another energisation is needed to cause the relay to return to the first rest condition.

Bitter Figure : The term used for a method of making magnetic domains visible.

Bitumen : A natural product having a complex mixture of hydrocarbons with inorganic materials, and melting at about 90 - 100°C. Refined bitumen finds use in the manufacture of insulating varnishes, impregnating compounds and joint-box compounds. Vulcanised bitumen finds use as a cable insulant.

Blocking Capacitor : It is a capacitor which is used in an electric circuit with pulsating current to disallow the passage of its direct component in part of the circuit.

Blow-out Coil : A device which is used for lengthening an arc in air by magnetic action, especially in d.c. contactors.

The coil has been mounted on a steel core between steel pole-pieces, and gets connected in series with the circuit to be broken. The current in the coil produces a magnetic field across the contact tips. When an arc gets formed a mechanical force gets produced on it, urging it upwards and away from the contracts. The arc consequently gets lengthened and rapidly extinguished, reducing contact wear and burning. A blow-out coil has been usually associated with an arc chute to prevent the arc from wandering.

Bobbin Insulator : Alternative name for shackle insulator.

Bobbin Winding : Transformer winding which has been used mainly for the highvoltage windings or small transformers, in which all the turns have been arranged on a bobbin.

Bohr Magneton : It is the unit of magnetic moment in atomic phenomena. Electron spin makes an electron-charge rotation on an axis, equivalent to a current loop magnetising along this axis having a magnetic moment. The value of the Bohr magneton has been 1.165×10^{-32} Wb m.

Bolometer : It is an instrument which is used for measuring radiant energy. The resistance of a fine wire or strip varies when it gets exposed to the radiation.

Booster : It is a device which is used to give a small increase to a normal-voltage supply, being connected in series with one line. D.C. boosters have been normally low-voltage d.c. generators which are used for adjusting a supply voltage, in line-drop compensation and as an aid in controlling the charging of large accumalator batteries.

Booster Transformer : It is device which is used for regulating the voltage on a power supply network. It finds use to add or subtract a voltage and is used at the sending end of a long h.v. radial feeder. It has been connected in series with the feeder as shown for one phase in Fig. B. 5., this transformer being supplied from another transformers have been connected in each of the three phases. The added voltage is in phase with the phase to-neutral voltage of the system, *i.e.*, it adds to it (or substracts from it) arithmetically. The magnitude of this added voltage can get varied in steps by varying the tappings on the secondary of the supply transformer.

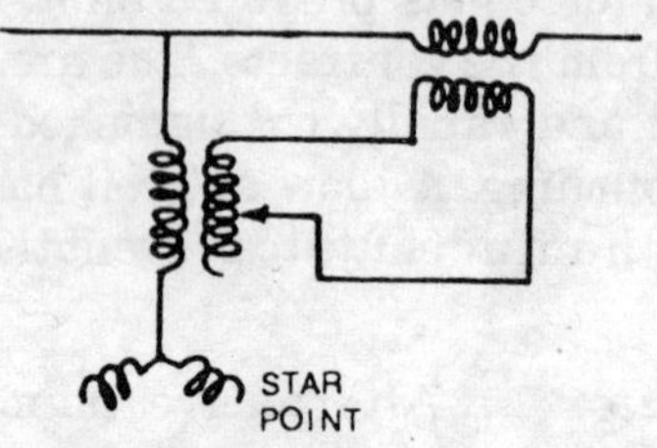

Fig. 5. Booster transformer. The connections are shown for one phase of a feeder.

Bow Collector : Refers to a sliding current-collector which is used in electric traction with overhead wires. It consists of a bow-shaped contact strip which gets mounted on a hinged frame.

Boys Camera : Camera having two lenses which are moving in opposite directions. It facilities the study of time characteristics of lightning and similar phenomena.

Braiding : Of a cable, a plaited protective covering,

Branch : Refers to a portion of an electric network. It is having two terminals and consists either of a single circuit element or a series-combination of two or more circuit elements.

Branch Joint : Refers to a form of cable joint where a connection to one continuous main cable has been done by another and, often smaller cable forming a branch connecting. It may be a breeches joint in which the branch cable has been taken parallel to the main cable and out at one end of a lead sleeve where a double plumb made to the cable and out at one end of a lead sleeve where a double plumb has been made to the cables, or a T-joint in which the branch cable has been taken away at right angles to the main cable and requires a pressed lead or copper box of suitable shape.

Break : Refers to the total minimum gap which gets introduced in one pole by the contacts of a switch or circuit-breaker when this has been in its fully open position.

Break Time : Of a circuit-breaker; see total break time.

Breakdown : See electric break down.

Breakdown Field Strength : Refers to the electric filed strength at which electric breakdown gets initiated.

Breaking Capacity : Of a circuit-breaker, see asymmetrical breaking capacity, symmetrical breaking capacity.

Breather : Desiccating device which has been fitted to conservator tank of a transformer to ensure that no moisture gets drawn in with inspired air.

Breeches Joint : Refers to a form of branch joint in which the branch cable has been taken away parallel to the main cable. It is also known as a trousers joint.

Bridge : Refers to a arrangement of resistors, capacitors, inductors, etc., usually in a four-sided circuit which is used for measuring an electric quantity by a null method. One of the diagonals gets connected to the current source and the other to a measuring instrument.

Bridge Transition : Of d.c. motors this term is used for a method of changing the connection from series to parallel without breaking the main circuit and while maintaining a similar current in all the motors.

Bridged-T Network : Refers to a two-port network having two-thermal circuit elements which get arranged similarly as the particular ones in the special type of attenuator.

Britannia Joint : Refers to a joint which is used for overhead transmission lines. The two lines to be joined have been bound together over a length of several inches.

Broad-base Tower : Refers to a lattice tower, each leg of which gets anchored separately.

Brush : A conductor which is used to make contact between a stationary and a moving surface. Operating conditions get varied over a wide range, and numerous types and grades of brush get developed over the years to match progress in machine design.

Brush Discharge : Refers to a form of intermittent gas conduction around a conductor when the electric field strength has been insufficient to cause a spark or an arc. It is having a feathery appearnce and is generally accompanied by a hissing or crackling noise.

Brush Shift : Refers to the movement of the brushes on a commutator away from the neutral position. In a d.c. machine without commutating poles a shift of the brush position may be made so as to improve commutation; in a repulsion or Schrage motor to vary the speed. It is also using brush lead.

Brushless Generator : A generator using silicon rectifiers as static commutation device. It has been of particular value as an aircraft generator, difficulties having been experienced having sliding contacts under conditions of high running speed, dry rarefied air and wide temperature range.

Buchholz Relay : Refers to a protective relay on a power transformer. An electric fault will always produce gas, and the relay has situated so that all gas released must pass it, *e.g.* in the pipe connection to the conservator

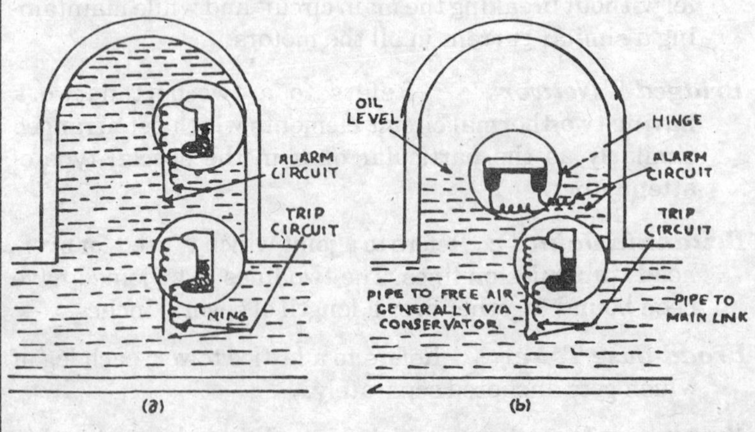

Figure 6. Buchholz relay. The normal condition has been shown in (a) with the relay full of oil. In (b) gas has formed and collected in the relay, the oil level has fallen and the first float has turned about its hinge allowing the mercury switch to close an alarm circuit.

vessel. The relay has been a chamber, normally full of oil, in which a hinged float carries a mercury switch which has been open when the float has been in its highest position. If gas has been formed it will collect in the chamber, the float will fall ahout its hinge and at a predetermined point the switch will close, and in common practice, produce an alarm signal. A second similar float and switch has been mounted on a hinge in the opening of the pipe connection from the transformer. It gets operated on the surge of oil up the pipe which will result from the rapid formation of gas due to a severe fault. The switch on the surge float has been invariably set to trip the circuit-breakers. (See Figure 6.) Alternative term for gas-and pressure actuated protective relay.

Buck : Effect of a booster which gets connected to reduce rather than increase the resultant voltage.

Buffer Battery : It is a battery which is included in a d.c. generation system and connected across the lines, to smooth out variations in supply voltage and current.

Bulk-oil Circuit-breaker : Refers to a circuit-breaker in which arcing occurs in oil, the contacts being enclosed in an oil-filled earthed tank. It has been sometimes referred to as a dead-tank circuit -breaker. circuit breakers of this type are available for a.c. systems for all voltages and repturing capacities up to 330 kV.

Bull Ring : In ovedrhead construction this term refers to a metal ring which is employed at the junction of three or more straining wires.

Bunch Filament : Refers to a lamp filament which is having two levels of support with the filament suspended between. It is bunched for compactness for employing in applications involving lens systems, where the focal point is critical.

Bunched Cables : Two or more cables which have been enclosed within a single conduit or way.

Bunched Conductor : Refers to a conductor which is composed of several wires twisted together in the same direction, and with the same lay throughout.

Bundle Conductor : Refers to a conductor which each phase conductor consists of two, three or more small stranded cables with a proper spacing maintained between them and a total section equal to that needed for current conduction. The bundle construction, although raising mechanical difficulties in erection and operation causes a significant reduction in the surface electric field intensity (see Figure 7.) Furthermore, it confines the higher in tensities to a comparatively small surface area, so reducing corona loss on two courts.

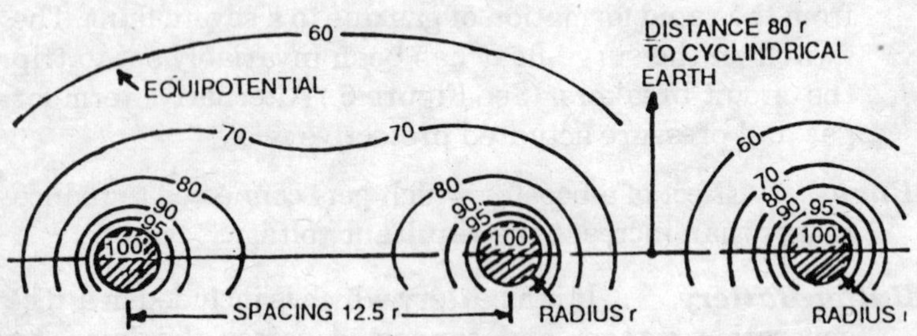

Figure 7. Bundle conductor. The equipotential field is composed with that of single

Burden : Refers to the load in volt-amperes on the secondary of an instrument transformer.

Bus Line : In a railway train, this term is used to refer to a cable (or train line) which is extending throughout the length of the train. It is used to interconnect collector shoes of similar polarity.

Busbar : Refers to a substantially equipotential conductor which is forming a terminal or junction point in a power system, for the connection of supplies and feeders.

Busbar Protection : Refers to a form of protection which automatically is able to disconnect a busbar system from its source of supply in the event of an insulation failure. Two basic methods have been frame fault protection and balanced current protection. The choice of scheme for a given application has been governed to some extent by economics. Schemes may be having varying complexity, either affording earth-fault protection only, or providing phase-to-phase fault protection as well, and covering either the individual busbar sections or the complete installation.

Bushing : It is an insulator which is generally cylindrical in shape and used for insulating and fixing a conductor where it passes through earthed structures like the walls separating parts of a X-ray equipment, or the tanks of power switches and transformers. It is also called lead-in insulator.

Butt Contacts : Refers to contacts which are brought together with a minimum of relative movement between fixed and moving components, relying mainly on the applied pressure for satisfactory conduction. Contact get established either at a point or along a line. The current generally gets interrupted at the point where normal conduction takes place.

Butt Seam Welding : Refers to a varied of butt welding in which roller electrodes have been used to conduct the current to the two sides of the joint and separate devices have been used to force the two edges together. Thereby forming a continuous well as in the case of tube formed from strip.

Butt Welding : Refers to a form of resistance welding in which the parts being joined have been butted together end to end or edge to edge. Varieties of butt welding include upset butt welding, flash butt welding and butt seam welding.

Buzzer : The term used for signalling device like an electric bell without hammer or gong. The sound have been produced by the electromagnetic vibration of an armature.

Cable : The term used for a length of one or more conductors which are laid up together, each conductor having its own insulation. The three main components of a cable have been the conductror, the insulation and the protection. The two principal insulating material most commonly used have been oil-impregnated paper and natural or synthetic vulcanised-rubber compounds, but in recent years plastic compounds like polyvinyl chloride and polyethylene finds extensive use in place of rubber. The protection for an impregnated-paper cable has been firstly, a lead or aluminium sheath to disallow ingress of moisture and, secondly, layers of bituminised-fabric tapes with or without a metal armour for protecting the sheath against corrosion and mechanical damage. For rubber- and plastic-insulated cables, the protection gets governed by actual service requirements.

Cable Bond : Electric connection for the metallic covering of a cable.

Cable Coupler : A device which consists of two engaging parts for connecting two sections of flexible cable. An interlock may get incorporated to disallow operation of the coupler when the circuit is live.

Cable Joining : The term used for the connection of one piece of cable to another in such a way that the conductivity and insulation at the joint have been equal in quality to that of the body of the cables concerned. The main types of joint used on normal distribution cables have been straight-through joint and the branch joint.

Cable Laying : Placing cables in position for the passage of electric power. Cables have been laid direct, *i.e.*, buried directly in the earth, or drawn in *i.e.*, drawn into conduits or ducts already laid.

Cadmium Cell : A standard cell which is alternatively known as a Western cell.

Cadmium Electrode : It is a rod of metallic cadmium within a perforated insulating tube, and provided with a flexible connection. By connecting through a suitable voltmeter to either the positive or the negative terminals of a lead-acid cell, the cadmium electrode when immersed in the electrolyte provides readings that reveal the condition of the cell plates.

Cadmium-nickel Cell : See nickel - cadmium cell.

Cadmium Test : Refers to a method of testing the condition of the plates in an accumulator cell. This test is involving the insertion of an electrode which is made of cadmium into the electrolyte near the middle of the cell and the measurement of voltage between the electrode and a terminal.

Cage Motor, Rotor, Winding : See squirrel-cage motor, squirrel-cage rotor, squirrel-cage winding.

Calculated Effective Area : Of the conductor of a cable, this term refers to the cross sectional area of a solid conductor of similar resistivity and having the same resistance as an equal length of the cable.

Calibration : Refers to the determination, by experimental observation, of the values (in appropriate units) of the readings of an instrument which is having an arbitrary or inaccurate scale.

Calomel-electrode System : It is a mercury electrode in a potassium-chloride solution which is saturated with calomel (mercurous chloride).

Cam : In an electric-lift system, cam refers to a wedge-shaped component fixed in the well or on the lift car or counterweight is able to operate the control apparatus when the lift moves.

Camera Tube : It is a cathode-ray tube having a photosensitive faceplate. It is designed to derive an electric signal by electric means from an optical image formed on the faceplate. It finds use in television.

Cap :

1. Refers to a metal fastening on an insulator by which it may be suspended and connected to another insulator in a string.
2. Refers to the terminal base of lamp.

Capacitance : The term used for the capability C of two conductors separated by an insulant to store an electric charge Q at a voltage V applied between them. In SI units C=Q/V, where C is in farads, Q in coulombs and V in volts.

Capacitor : Refers to a circuit element, the predominant characteristic of which has been capacitance, and which stores energy in its electric field. A practical capacitor will be having energy loss in dielectric hysteresis and conductivity, and could be represented by a pure capacitor with a series or a shunt loss-resistor.

Capacitor Motor : It is a single-phase machine having a split stator winding, with an external cpacitor in the auxiliary circuit for increasing starting torque and decrease starting current. The capacitor gives phase displacement so that the current in the auxiliary winding leads that in the main winding. It could be used during the starting period only (capacitor-start motor), or get permanently connected. Alternatively, a motor may start with one value of capacitor and run with a different value (capacitor-start-run motor).

Capacitor Transformer : Refers to voltage transformer for transmission lines which is operating at 100 kV and upwards. It consists of two capacitors connected in series across line and earth, so as to provide a voltage divider effect. A burden gets connected through a step-down magnetic transformer with an appropriate leakage reactance (see Fig.1). As an alternative to the magnetic voltage transformer it is having lower cost and greater service reliability, and can also be used as a connection element in carrier-frequency protection.

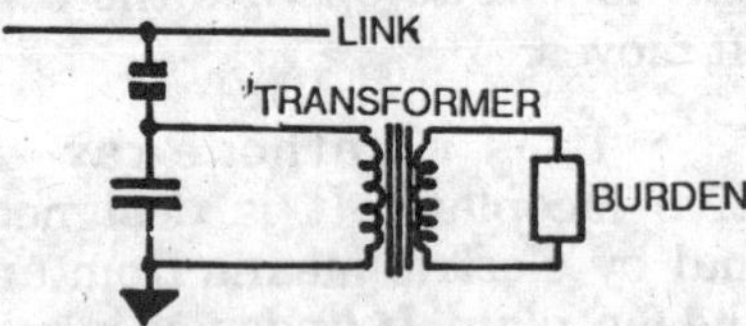

Figure 1. Capacitor voltage transformer.

Capacity : Refers to the quantity of electric charge which may be taken from an electric cell or battery or an accumulator at a given rate of discharge. It is generally expressed in ampere hours.

Carbon : Non-metallic element of adjustable conductivity. It has been highly refractory, and as its oxides are gaseous, no surface films are produced. It has been chemically resistant and self-lubricant. Its contact resistace varies inversely with the pressure. Carbon finds use in various forms for brushes (see carbon brush, electrographitic brush, etc.), for arc electrodes (see carbon arc lamp, carbon arc welding) for cell electrodes (see Leclanche cell), for contacts, for current collectors in electric traction, for pressure-sensitive resistors in microphones, carbon-pole regulators, etc., and as a furnace material.

Carbon Arc Lamp : It is a lamp which is consisting basically of two carbon electrodes between which an elecrric arc is maintained. The carbon arc is used as a source of light for general illumination, but its high brightness, compact source and colour find valid application in film projection and searchlights.

Carbon Arc Welding : It is a form of non-consumable electrode arc welding in which the arc gets struck between a carbon rod and the workpiece. The quality of the weld has been low owing to inadequate protection of the weld pool from atmospheric contamination.

Carbon Brush : It is one of the first types of brush to get developed to supersede the metal wire or foil brushes used on the earlier electric machines. The raw material, amorphous carbon in various forms such as lamp black, gas carbon or coal, has been first reduced to a very fine powder. It has been then mixed with a suitable binding material such as pitch or tar and compressed into blocks which are baked under controlled conditions at a fairly high temperature to carbonise the whole mass of the material and to impart the required mechanical and electric characteristics.

Carbon-pile Regulator : In the simplest form, refers to an assembly of carbon plates with means for compressing the assembly, usually a simple handwheel and screw mechanism. The resistance of the stack of plates get varied with the pressure, giving a convenient method of controlling currents up to the order of 100 A.

Some types of automatic voltage regulator use a carbonpile rheostat.

Another type of carbon pile rheostatic regulator is illustrated in Fig. C. 2.

Carey-foster Bridge : It is an adaptation of the Wheatstone bridge which gives a greater accuracy with resistances of low and medium value. It is especially useful for getting the divergence between a nominal-valued resistor and a supposedly identical standard. A feature has been the use of a low-resistance slide-wire of length I and resistace r per unit length. The ratio arms P and Q have been nominally equal, and the unknown R is nominally equal, and the unknown R is nominally equal to S. Balance is obtained with the conditions of Fig. 3 so that R

$$\frac{P}{Q} = \frac{R+ar}{S+(1-a)r} = \frac{R+S+1r}{S+(1-a)r}-1$$

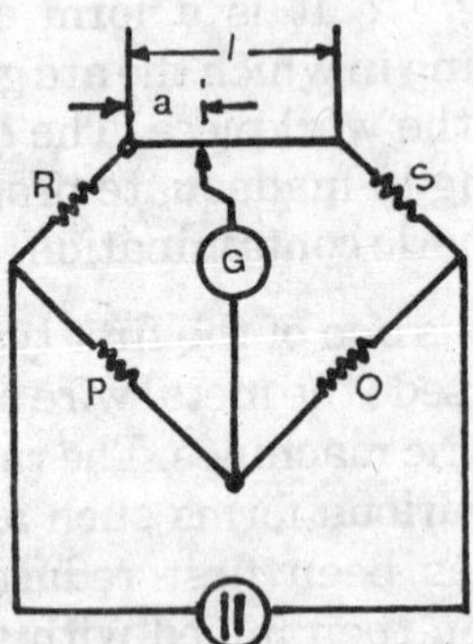

Fig. 2. Carey - Foster bridge

and S have been now interchanged, and balance obtained at a setting b. Then, we have

$$\frac{P}{Q} = \frac{R+S+1r}{R+(1-b)r}-1$$

whence

$$S\text{-}R=(a\text{-}b)r$$

Carrier : It is a sinusoidal voltage or current which is used for signalling by line or ratio. Intelligence has been impressed on it by a process of modulation.

Carrier Current Protection : Refers to a form of protection for power transmission lines which uses a high frequency signal transmitted over the line itself. Fig. 3 shows the essential components at one end of a line. The transmitter and receiver have been connected in parallel and feed the line-coupling equipment via a low-loss cable. Connection to the line has been through two highvoltage capacitors CC; they have been series-tuned by h.f. inductors LL to accept the carrier frequency. A path for the 50 Hz current of the capacitors has been provided by a parallel-tuned circuit having a high impedance at the carrier frequency. The line traps have been parallel-tuned circuits which reject the carrier frequency; the signal has been thereby confined to the protected line.

There have been three principal types of carrier system in common use, also a scheme of carrier intertipping. All are using fault detecting or 'starting' euipment to avoid the necessity of continuous high-level carrier transnmission.

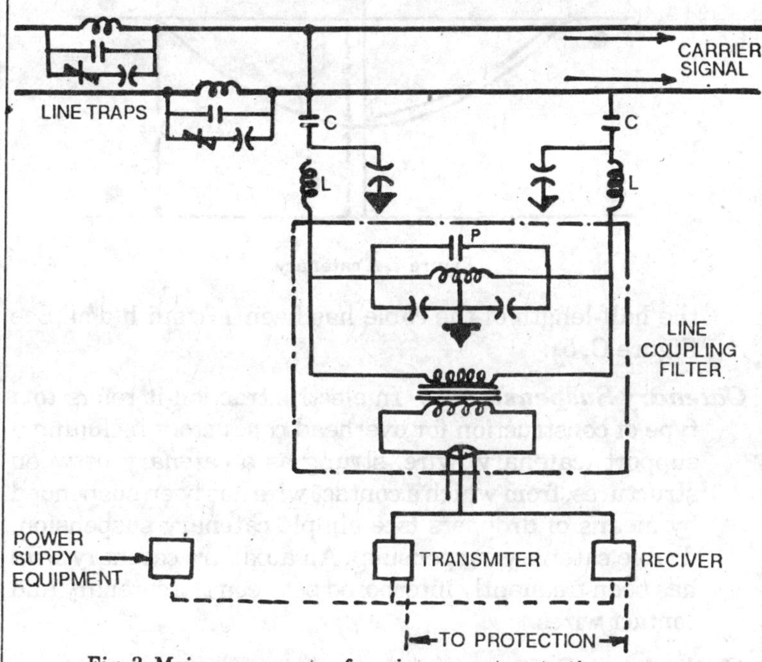

Fig. 3. Main components of carrier current protection system.

Carter Coefficient : A numerical coefficient which is used in the analysis of the magnetic flux distribution in the air-gap of an electric machine. It depends upon the effect on the distribution of the presence of slots or ducts.

Cartridge Fuse : Refers to a fuse in which the fuse element has been totally enclosed in a cartridge of insulating material. It is usually tabular with matallic end-caps.

Cascade Connection : See concatenation.

Castell Interlocking : It is a method of ensuring that switchgear and other equipment has been operated in the correct sequence or under the right conditions.

Catenary : Refers to the curve which is taken up by a chain (or a cable) of uniform weight per unit length when freely supported from two points. Consider a cable of length 2. The catenary equation has been given by the expression y=c cosh (x/c). The sag has been $s=Y_1-c$, and

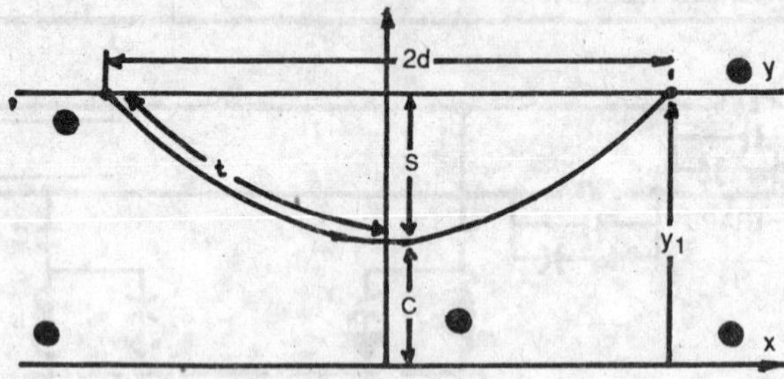

Figure 4. A catenary.

the half-length/of the cable has been 1=c sin h(d/c). See Figure C. 5_4.

Catenary Suspension : In electric traction it refers to a type of construction for overhead conductors including a support (catenary) wire, strung as a catenary between structures, from which a contact wire has been suspended by means of droppers (see simple catenary suspension, double catenary suspension). An auxiliary catenary wire has been frequently interposed between the catenary and contact wires.

Cathode : Refers to that electrode (*e.g.* in an electron tube, mercury are rectifier, electrolytic bath, etc.) to which the current flows from the anode. The current direction has been taken conventionally as that of the movement of positive charges. In an electronic vacuum tube the current is, in fact, solely a transfer of electrons emitted from the cathode.

Cathode Drop : Refers to the difference in voltage, in an electric welding machine, between that at the negative end of the arc stream and that of the adjoining conductor.

Cathode-ray Tube : Refers to an electronic device which includ an electron gun with means to produce an intensity-controllable filamentary, focused electron beam within a highly evacuated envelope that has opposite the electron gun a round or rectangular faceplate with a thin inside coating of phosphor for cathodoluminescence. The position of the luminous spot produced on the faceplate has been variable and determined by means provided to control precisely the transverse deflection of the electron beam. Such a tube having appropriate design has been as an oscilloscope or for an oscillograph, or in television, radar, and data or information display equipment.

Cathodic :

(1) Having a more negative electrode potential.

(2) Indicating an element below hydrogen in the electrochemical scale.

Cathodic Protection : The term used for defence against corrosion which can be applied to cables drawn into ducts or buried direct. It is dependent on the fact that electrochemical corrosion results from the existence of potential difference between metals and the electrolyte surrounding them. The corrosion tends to occur at the anode.

The technique is making the potential of the sheath and armour more negative to the electrolyte (water or soil) than a limiting value (about-550 mV for lead). All parts of the metal then become cathodic. This result can be realised by connecting to the cable sheath reactive anodes composed of an alloy of base metals such as magnesium or zinc.

Cathodoluminescence : This term used for the excitation of luminescence in a phosphor by the impact of electrons from an external source. Certain phosphors can get excited with an electron (or cathode ray) beam of 100

V or less; this is low-voltage cathodoluminescence. On the other hand, the accelerating voltage in a conventional television picture tube has been about 15 kV, producing an image by high-voltage cathodoluminescence.

Cation : A positively charged ion, existing in an electrolyte or gas conduction. It will travel towards the negative electrode or cathode under the influence of an applied voltage.

Catolyte (or catholyte) : Refers to the portion of an electrolyte surrounding the cathode, which has been affected chemically by cathodic reactions.

Cavity Resonator : It is a device which formed by closing a length of waveguide at both ends, so that a wave of appropriate frequency gets excited in it. Standing waves will be set up due to reflections from the ends, and the device becomes a resonator.

C-core : See cut-wound core.

Cell :

(1) refers to a chemical source of electric energy which is consisting of two electrodes in an electrolyte.
(2) Compartment of cellular switchgear.

Cellular Switchgear : It is a form of switchgear in which the components have been housed in individual cells. Each circuit is having a separate cell divided into compartments for circuit-breakers, busbars, isolators, voltage transformers and cable sealing boxes. Each compartment is having its metal access door, which may get interlocked with other equipment to ensure the safety of maintenance personnel. The cells are often made of moulded stone, carried in a steel framework.

Cellulose Acetate : It is a thermoplastic which is available as sheet, film, moulding powder, extruded sections or lacquer solutions. It finds use for a number of applications in the electrical field. Typical applications of film have been wire and coil inter-layer insulation, while mouldings are used for such components as terminal blocks and switch dollies. In sheet form it has been used for instrument dials.

Centre-contact Cap : Refers to a bayonet cap in which the shell is coming one contact, the other consisting of a central projection. It may be referred to as SCC (small centre contact).

Ceramic Insulators : Insulators which are made of ceramics. They find use in high-voltage power transmission, switchgear, transformer bushings and fuse-holders because of their good mechanical strength, weathering resistance, high electric strength and resistance to arcing In aerials, electron tubes, resistor bodies and coil formers they have been useful due to high electric resistivity, and dimensional and thermal stability. The Table summaries their properties.

Ceramic material	*Relative permittivity* $\in r$	*Loss power-factor tan s × 104*	*properties*
Porcelain	5 - 7	50 - 200	Plastic for forming
Steatite	6	<10	Low loss
Forsterite	6	4	High thermal expansion
Cordierite	4.5	40-100	Low thermal expansion
Alumina	8.5	3 - 9	Gas-tight

Ceramics : Articles which are formed from substances having clay and converted into hard brittle bodies by a high-temperature treatment, or other products such as silica refractories that are made by a similar process. They have been used to a large extent in the manufacture of insulators.

Chain Winding : Alternative name for basket winding.

Change-of-linkage Law : It is a descriptive term for Faraday's law of electromagnetic induction. If there has been any closed linear path in space, and the magnetic flux ϕ surrounded by the path varies with time, then an electric field appears in the path, summing in the closen path to

$$e = -\partial f/\partial t$$

The negative sign means that the direction of the electromotive force has been such as to produce a current opposing the change flux.

Change-pole Motor : Refers to an induction motor in which the speed gets altered by changing the number of primary (stator) poles.

Characteristic Impedance : It is a function of a transmission line. The propagation of a voltage disturbance v along a line gets accompanied inevitably by a current i, because the travel of the voltage implies the charging of the line capacitance C, for which i has been necessary. Similarly the travel of the current set up magnetic flux in accordance with the line inductance L. The ratio $v/i=Z_0$ has been the characteristic impedance, and in a loss-free line $Z_0=\sqrt{(L/C)}$, where L and C has been expressed per unit length of line. Abbreviation: c.i.r.

Charge :

1. Of a body or a system, the difference between the numbers of positive and negative elementary charges contained in the body or system. The charge would be positive when the difference has been positive, and negative when the difference has been negative. The SI unit has been the coulomb (symbol: C).
2. Of a capacitor, the quantity of charge on its positive plate.
3. Of an accumulator, the time-integral of current supplied to restore its plates to a state in which it can supply electric energy. This is usually measured in ampere-hours.

Charge Carrier : An elementary particle such as an electron which represents a single negative elementary charge, a proton and a hole each representing a single positive elementary charge and an ion having one or more elementary charges which are either positive or negative. The magnitude of an elementary charge has been 0.1602×10^{-18} coulomb.

Charge Indicator : It is a visual indicator of the state of charge of a lead-acid cell in terms of its specific gravity. A direct measure of the specific gravity has been indicated by a hydrometer, a hollow-glass bottom-weighted tube which floats in the electrolyte and is marked with an appropriate scale. Pilot-ball indicators have been wax or hollow glass spheres that will sink or float in accordance with the density of the electrolyte.

Charging Current : Refers to the current which is taken by a capacitor when its voltage difference is being changed.

Charging Resistor : Refers to a resistor which is connected in circuit with a circuitbreaker or switch to prevent the current rising at an excessive rate when the breaker or switch is closed.

Chassis : Refers to a conductive body which is usually made of sheet metal, on which electric components have been assembled and whose potential has been taken as reference.

Choke : Deprecated term for inductor.

Chopped Wave : Refers to a wave in which the voltage decreases abruptly from near maximum to zero. It is used to determine the ability of a transformer to withstand the effects of high unidirectional voltages resembling lightning surges, it could be obtained by paralleling a surge gnerator with a rod-type spark gap, which has a delayed breakdown. The chop simulates the sudden disappearance of lightning surge consequent upon spill-over or the action of a lightning arrester in service.

Chopping : The term refers to the action of a circuit-breaker in interrupting an alternating current prior to a zero-pause.

Choreded Coil : Refers to a distributed winding in which the coil span has been less than the pole-pitch. It is also known as a short-pitch coil.

Chording : Reduction (or occasionally, the increase) of the span of a coil in an electric machine compared with full pitch (equivalent to a pole pitch).

Chromel Nickel : It is a chromium alloy which is sometimes containing iron. It finds use in thermocouples, for heating elements and in heat resistance applications.

C.I.R. : Characteristic impedance ratio.

C.M.L. : See continuous maximum load.

C.T. : See current transformer.

Circle Diagram : A term which is usually understood to mean the polar diagram of the stator current of an induction motor, but actually descriptive of the current in several other kinds of a.c. machine.

Circuit : The term used for the closed loop of a solenoidal function; the endless path around which either in reality or by analogy, a physical quantity flows; in particular the path of an electric current or of a magnetic flux.

Circuit Analysis : See network analysis.

Circuit-breaker : A device which is used for automatically making and breaking the normal current in a circuit or the current which may flow under fault conditions such as short-circuit. Circuit-breaking has been effected by the mechanical separation of contacts connected in series with the circuit, in an insulant medium which helps in extinction of the arc so formed and which restores the electric strength of the break. Insulants in common use have been air and oil, although other media have been possible and water has been used successfully. Miniature circuit-breakers have been replacing fuses in domestic instalations.

Circular Mil : Unit of area which is equal to the area of a circle whose diameter has been one mil ($=2.5\pi\times10^{-7}in^2$).

Clamping Die : Of a resistance welding machine, alternatively known as a contact jaw.

Clark Cell : An early form of standard cell which is developing 1.433 V at 15°C. One electrode has been of zinc in zinc sulphate the other of mercury in mercurous sulphate.

Class A, B and C Operation : Refers to conditions under which a thermionic valve could be operated as an amplifier or oscillator. The conditions could be determined by the amount of negative grid-bias applied, in accordance with which the valve introduces varying amounts of distortion.

Cleat : Refers to an insulated incombustible cable support. It is generally used for vulcanised-rubber-insulated, taped and braided or p.v.c. insulated cable.

Clock-hour Figure : Of a transformer winding, this term refers to the method of using the numbers 0 to 12 as on a clockface, with 12 meaning 360°, for designating the phase displacement of the no-load voltage phasors at the terminals of a given low-voltage winding and the associated high-voltage winding.

Closed Circuit : Refers to an electric circuit which provides an uninterrupted path for a current.

Closed-circuit Alarm System : Refers to burglar alarm system in which the alarm gets sounded if wires are broken, in distinction from an open-circuit system in which the alarm has been given when the circuit has been closed.

Closed-circuit Television : A type of television system in which the signal has been fed from the camera to the receiver by cable and not by radio transmission.

Closed loop Control : Refers to a system of control of a machine or process in which the result of an input control setting has been compared with the input, and any difference (error) has been used to correct the result (output) to bring the error towards zero.

Coaxial Cable : Refers to a cable which is suitable for transmission of radio-frequency signals with frequencies about the v.h.f. range. It occupies a place which is intermediate between an open-wire feeder and a waveguide. A central conductor gets separated from a surrounding coaxial conductor by a solid insulant like polythene, or by air. In the latter case, the separation has been maintained by a series of disks.

Cockroft-Walton Multiplier : Refers to a voltage multiplier which is having two banks of series capacitors alternately connected by rectifiers acting as changeover switches according to the instantaneous output polarity of the energising transformer. A machine of this type finds use to inject protons into a larger particle accelerator such as a proton linear acceterator. See Fig. 5.

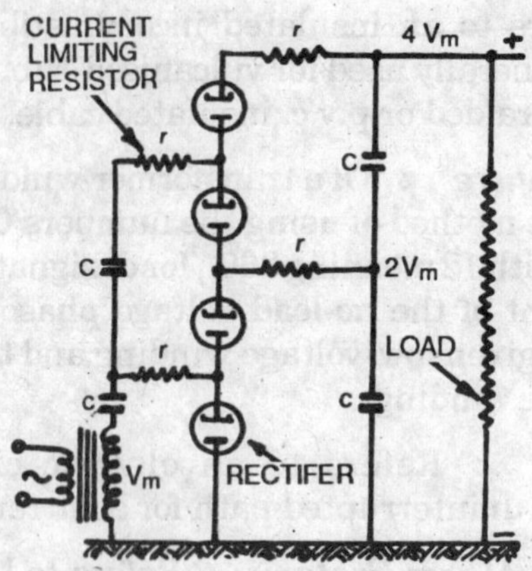

Fig. 5. Two stage four-valve Cockroft-Walton high-voltage d.c. generator.v_m=transformer peak output voltage.

Coefficient of Coupling : Refers to ratio of the mutual impedance components (inductive, capacitive or resistive) of two circuits to the square root of the products of their total impedances of the same kind, *i.e.* inductive, capacitive or resistive. That is, $K = xm\sqrt{(x_1 x_2)}$ where x_m denotes the common mutual reactance; and x_1 x_2 denote the total similar reactances of the primary and secondary respectively taken separately. The coefficient expresses the degree of closeness with which the primary and secondary are coupled.

Coefficient of Mutual Induction : See mutual inductance.

Coefficient of Self-induction : See self-inductance.

Coefficient of Utilisation : Refers to the proportion of light which is emitted by a lamp which actually reaches a working plane.

Coercivity : See B/H curve.

Coil Ignition : It is a system of h.t. supply in automobiles. The interruption of the current from a battery makes a high secondary electromotive force to be induced in a coil winding for distribution to the sparking plugs.

Coil Span : Refers to the distance between the two sides of an armature coil. It has been measured round the periphery of the armature and stated, usually, in electric degree or slots.

Cold-cathode Emission : Emission which is occurring from an unheated cathode. When an electric field intensity of the order of 10°-10^{10} V/m has been applied to a metal surfance and has been so directed as to accelerate electrons from the surface, emission will occur even at low temperature. This is also termed as field emission.

Cold-cathode Lamp : It is a form of long-life fluorescent lamp in which the electrodes operate at about 200°C and have been in the form of shells of high-purity steel housed at the ends of the tube.

Cold-cathode Rectifier : It is a device which is dependent on the asymmetrical current/voltage characteristic of a tube having an oxide-coated electrode of large area and a smaller, uncoated electrode; the former is the cathode and the latter the anode. The difference in work-function of the two kinds of electrode surface makes field emission possible from the cathode at compartively low voltage gradients.

Collective Control : Refers to a method of automatic lift control. Calls from the lift-car and landings has been stored, and the lift-car stops in floor sequence at each landing from which a call is made. In directional collective control, there have been provided up and down buttons, and the lift-car deals with all calls in one direction then reverses for the remainder.

Collector Ring : Alternative term for slip-ring.

Collector Shoe : In electric traction, it refers to a metal device which is kept in sliding contact with the conductor rail.

Colour-matching Tube : Fluorescent tube which is coated with a phosphor to provide a colour temperature of 6500°K. This approximates to the light from an overcast north sky.

Commissioning : Testing an electric device or plant before it gets connected to a supply system to ensure as far as possible that it is in a sound condition and will make no disturbance when switched on. These tests should be designed to prove both the mechanical and the electric condition of the test object. Their number and complexity on the size and importance of the plant under consideration.

Commutating Pole : See interpole.

Commutation : Refers to the periodic transfer of current from one conductor to another in a mechanical manner as in a d.c. machine or in an electric manner as in a static converter.

Commutator : It is a essential part of all d.c. machines. It comprises of a number of copper bars which are insulated from each other and from earth and suitably supported to withstand the stresses due to rotation. Each bar gets connected to one end of at least two armature coils. Stationary carbon brushes would rub on the surface of the commutator and pass current between the winding and the terminals of the machine.

Commutator Motor : It is a motor having an armature which is incorporating a commutator. Single-phase commutator motors may be series or repulsion. A three-phase commutator motor has been a variable-speed machine in which an auxiliary voltage supplements the current flow in the armature. A commutator machine could get cascaded with an induction motor for controlling the speed of the latter. With motors of moderate or large outputs (above about 100 kW) the commutator machine has been a separate unit (Scherbius system) but for smaller outputs it has been integral with the induction motor (Schrage motor).

Compensated Induction Motor : Refers to a induction motor in which the excitation has been supplied to the secondary circuits at slip frequency by using either an external exciter or an additional winding, with commutator and brushes, incorporated in the motor. By suitable adjustment of the phase of the exciting current, unity power factor could be obtained in the primary circuits.

Compensated Voltmeter : It is an instrument which has been designed to indicate the voltage at a remote point. It is compensated for the voltage drop in the circuit between the remote point and its point of connection.

Compensating Winding : It is a winding on a machine which has been reduce some of the effects of armature reaction.

Compensation Theorem : It is a theorem which is used in network analysis. For any given set of circuit conditions, any impdedance Z in a network, carrying current, I could be replaced in the network by a generator of zero internal impedance and electromotive force E=IZ; that is, an e.m.f. directed against the current. An extension of the theorem may be stated that if an impedance Z be changed by +ΔZ, the effect on all other branche currents has been that which would get produced by an e.m.f. of –IΔZ in series with the changed branch. The theorem which valid for linear systems.

Compensator : Refers to a large shunt capacitor which gets inserted at major high voltage substations to improve power factor. Such compensation can be by banks of static capacitors or by a synchronous compensator.

Complex Permeability : With the assumption that magnetic flux density and magnetic field strength are sinusoidal, with B and H their respective phasors, the effects of hysteresis and eddy currents may be represented by the relation.

$$\mu B/H=\mu'-j\mu''$$

where μ is the complex permeability, μ′ corresponds to absolute permeability and μ″ represents the loss in the magnetic material.

Complex Permittivity : Refers to a consequence of the hysteresis and residual conductivity effects in an insulant. Under alternating stress there has been an effective conductivity σ which is including both ohmic and hysteretic effects. The total current density J in a material in an electric field of intensity E at angular frequency ω may be consequently

$$J=\sigma E+j\omega\in E=E(\sigma+j\omega\in)$$

$$=j\omega\varepsilon E$$

where ∈ is known as the complex permittivity, given by

$$\in = \varepsilon - j(\sigma/\omega)$$

The 'real' part has been the absolute permittivity, and the quadrature term has been associated with the loss component of the current density. The complex permittivity makes the straightforward relation

$$J = j\omega\varepsilon\in E$$

to be given, where the complex permittivity has been including the angle arctan $(\sigma/\omega\in)$ called the loss angle.

Complex Power : The complex quantity

$$S = S \exp j\phi = \varsigma I \chi o\sigma\ \phi + jVI \sin\phi = P + jQ$$

$$\text{where } V = V \exp j\alpha_v = V(\cos\alpha_v + j\sin\alpha_v$$

$$I = I \exp j\alpha_i = I(\cos\alpha_i + j\sin\alpha_i)$$

$$\phi = \alpha_v - \alpha_i$$

have been the phasors having r.m.s. values of sinusoidal voltage and current in an electric circuit. The quantity S refers to the apparent power, P the active power and Q the reactive power.

Complexor : Refers to a complex quantity which represents the quotient of two phasors of a different kind. Thus with the voltage phasor V and associated current phasor I in a circuit, the quotient V/I impedance complexor and I/V means admittance complexor.

Compound Catenary Suspension : In electric traction, this refers to a form of construction for overhead conductors which are designed to eliminate the 'hard spots' which occur in simple catenary suspension where the contract wire is attached to the supporting structure. The contact wire has been suspended from an auxiliary catenary by solid-wire loop droppers which have been free to slide vertically over the auxiliary catenary, which has been itself suspended by rigid droppers from the main catenary.

Compound-filled Switchgear : It is a switchgear which is having the space between the metal casing and the

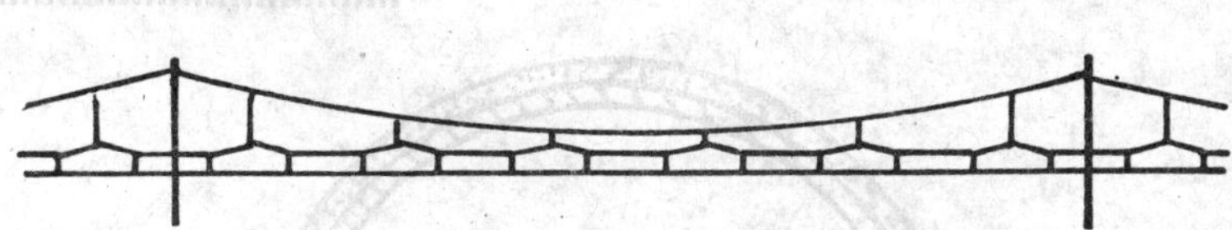

Fig. 6. Compound catenary suspension.

the conductive parts filled with an inclulating compund.

Compound-wound Motor : It is a d.c. motor in which the load characteristic of a shunt-wound motor has been tailored to requirements by a winding on each main pole connected in series with the armature circuit. If this series winding has been of such a polarity as to assist the main poles it has been termed as cumulatively compound-wound and if in opposition to the main poles as differenetially compound-wound or counter compound-wound or reverse compound-wound. The effect on the speed characteristic has been deficted. Figure 7.

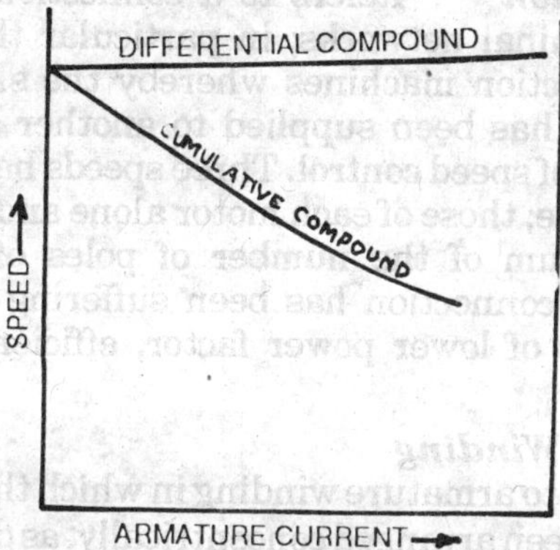

Figure 7. Load characteristics of differentially compound-wound and cumulatively compound-wound motors.

Compounding : This term used for a method of adjusting the characteristics, usually of an electric machine, in accordance with its loading.

Compton Electrometer : It is a sensitive quandrant electrometer, having a mirror on the moving system used with a fixed lamp and scale.

Compression Cable : It i san impregnated load-or polyethylene-sheathed single-or three-core cable , in which the conductor section has been oval, and the sheath has been thinner than normal (see Figure 8). The cable is included in a pressure container that gets charged with nitrogen gas at about 1.4 MN/m^2.

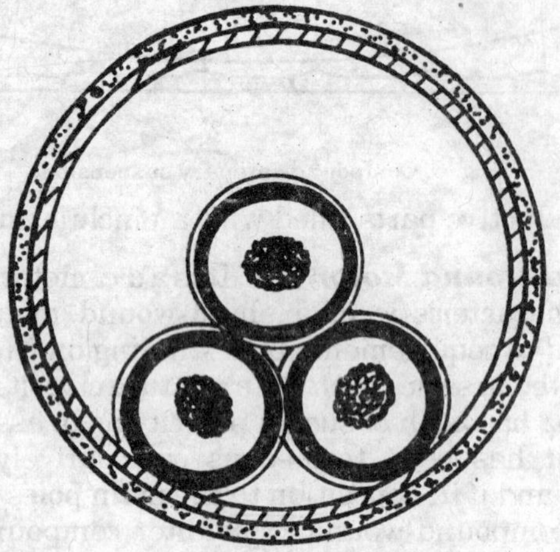

Figure 8. compression cable.

Concatenation : Refers to a connection in sequence of four-terminal networks; in particular the connection of two induction machines whereby the slip power of one machine has been supplied to another as input for the purpose of speed control. Three speeds have been normally possible; those of each motor alone and one determined by the sum of the number of poles of each machine. Cascade connection has been suffering from the disadvantages of lower power factor, efficiency and pull-out torque.

Concentric Winding

(1) Refers to armature winding in which the coils of a group have been arranged concentrically, as depcted in Figure 9.

(2) Transformer winding in which the primary winding has been arranged concentrically with the secondary winding.

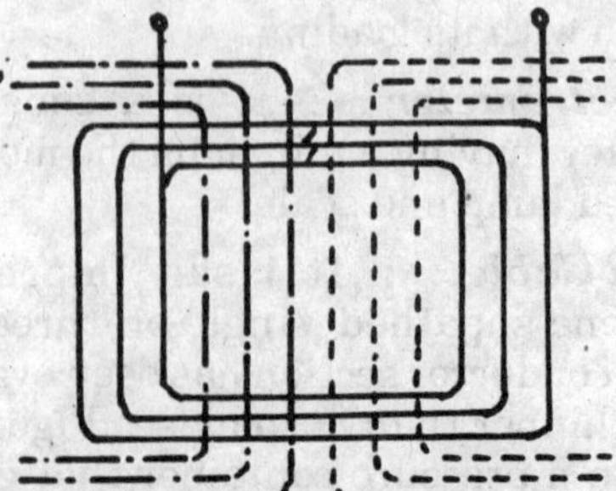

Fig. 9. Single-layer concentric winding.

Condenser : Obsolete term for capacitor.

Conductance : Refers to the dissipative part of admittance. In the case of steady-state d.c. circuits, the conductance G=1/R. The SI unit has been the siemens (Symbol S).

Conduction : Refers to the process by which an electric field has been enabled to maintain an electric current.

Conduction Pump : It is a type of electromagnetic pump which is having an external current source. In the d.c. pump shown in Figure 10 the current has been supplied to the liquid via busbars brazed to the tubewall; a few turns of busbar give the magnetomotive force for the magnetic field. Alternatively, a permanent magnet may be used, though it has been usually economical only in small pumps.

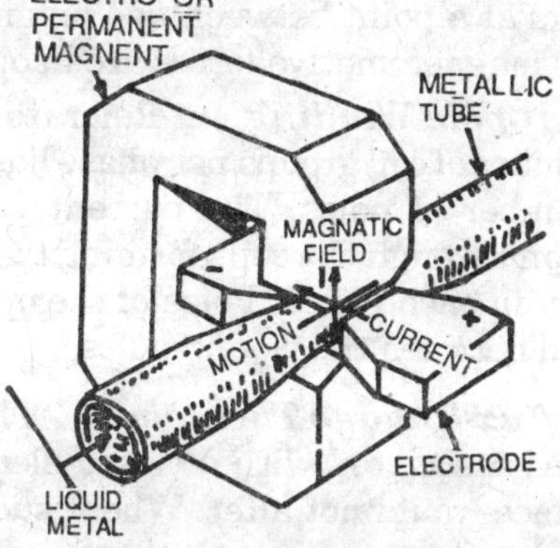

Fig. 10. D.C. conduction pump, used mainly with heavy metals such as lead, bismuth or mercury.

Conductivity : If specific property of a material which describes its ability to conduct an electric current. It refers to the conductance in siemens of a conductor of unit length and unit cross-section, under specified conditions (*e.g.* of temperature).

Conductivity Cell : In electrochemistry, a cell which is used for measuring the resistance of an electrolyte.

Conductor :

(1) Refers to material that provides a low resistance to the passage of electric current.

(2) Of a cable, the electrically continuous wire or wires

which form the conduction portion. They are generally composed of high-conductivity annealed copper, and with the exception of very small sizes, have been made up of a number of copper wires stranded together for flexibility.

Conduit : Trough or pipe which is used to contain and protect cables or wires. It can be of plain or screwed steel, earthenware, bituminised paper, fibre, plastics, etc., and be flexible or rigid.

Connector : A component which is used to interconnect two single-or multiple-conductor systems.

Consequent Pole

(1) A pole which is occurring on a part of a permanent magnet other than a free end.

(2) A magnetic pole in an electromagnetic circuit which occurs at a point between two magnetising coils when their magnetomotive forces are in opposition.

Consequent-pole Winding : Refers to a winding in which the number of coil groups per phase has been equal to half the number of poles. The current passes in a similar direction through the coils forming two poles of the same polarity in each phase. Poles of the opposite polarity get induced between the wound poles.

Constant-flux-linkage Theorem : The magnetic flux linkages associated with a closed electric circuit of zero resistance would not alter. Where such a lossless closed circuit has been carrying a current I and has been in consequence associated with a magnetic flux, it has magnetic energy amounting to $\frac{1}{2}$ LI^2, where L refers to the effective inductance. If the circuit is having no dissipative element, the stored energy must get associated with the circuit *i.e.* the current must remain constant.

Constantan : Copper-nickel alloy having a low temperature/resistance coefficient. It finds use in thermocouples and as the resistance wire in resistance boxes.

Contact Bar : Of a resistance seam-welding machine it is a device to exert pressure on the components being welded, and conduct the current to them. It is also known as an electrode bar.

Contact E.M.F. : The term used for the electromotive force which is occurring when two dissimilar metals are brought into contact. Free electrons diffuse in each direction across the boundary, and initially more electrons would pass in one direction than in the other, so that one metal would become positive to the other by a small contact potential. A condition of equilibrium gets reached in which the net electron transfer has been zero. Two or more contact e.m.f. tending to produce a direct current, having energy derived from a chemical source, form the basis of primary and secondary cells.

Contact Jaw : Of a resistance welding machine, this term refers to a device that holds the components being welded, and conducts the current to them. It is also known as a clamping die.

Contact Rectifier : It is a device which uses mechanically operated switches to convert alternating to direct current. The process of rectification has been similar to that of any other polyphase rectifier, and similar transformer connections can be used as for single-anode or ignitron valves.

Contact Wheel : Of a resistance seam-welding machine, it refers to a rotatable wheel which exerts pressure on the components being welded, and conducts the current to them. It is also known as an electrode wheel.

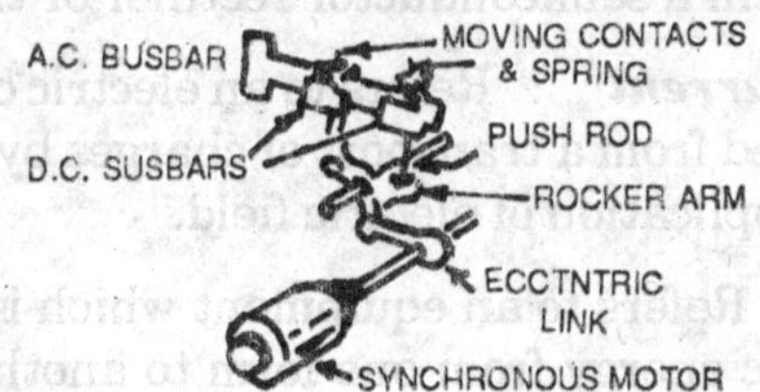

Fig. 11. Typical mechanism for contact rectifier.

Contactor : It is a device which is used for repeatedly opening and closing electric circuits. Electrically, relays and circuit-breakers also fulfil the same function. The difference between these and contactors is one of detail design.

Continuous Discharge Current : In a lightning flash. This refers to the relatively low current which is flowing continuously and on which much larger currents are superimposed.

Continuous Duty : Refers to the operating condition in which a device or equipment have been subjected to a substantial load that remains constant for an unlimited duration.

Continuous Maximum Load : Refers to the maximum load at which a device or equipment could be operated for an unlimited period under the specified conditions.

Control Board : A switchboard which incorporates indicating and operating devices and accessories for the remote control of switchgear.

Control Gear : Usually in heavy-current installations, if refers to the combination of switching devices with their associated equipments for control, measurement, protection and regulation.

Control Ratio : Of a thyratron, which is the ratio between the critical grid voltage and the related positive anode voltage.

Controlled Rectifier : A rectifier which is incorporating a means of regulating its output voltage. It will be a grid in the case of a mercury-arc rectifier or an additional layer of material in a semiconductor rectifier or thyristor.

Convection Current : Refers to an electric current which gets resulted from a transport of charges by means other than the application of electric field.

Converter : Refers to an equipment which is used to convert electric energy from one form to another, such as a rectifier, which converts a.c. to d.c. energy, an inverter, which converts d.c. to a.c. energy, a frequency converter which is able to convert a.c. energy from one frequency to another, and a phase converter to change the number of phases without change of frequency.

Converting Station : A substation which has been designed for the operation of converters.

Coolant : It is a material which is used to remove heat from an electric device or equipment. Four common substances have been air, hydrogen, oil and water. Air and oil have been by far the most common coolants because of their cheapness and insulating properties respectively, hydrogen is having the merit of a low density which causes a low windage loss for rotating machines, and also a high specific heat which provides it a heat-removal capacity similar to that of air.

Collant material	*Air*	*Hydrogen*	*Water*	*Oil*
Density, g/cm^3	0.0013	0.00009	1.0	0.83
Specific heat, J/g°C	0.84	14	4.2	2.2
Volume required to remove 1 kW at 150°C, m^3/min	3.4	3.4	0.00095	0.0025

Copper : It is having high electric and thermal conductivity. It is good all-round mechanical properties, good resistance to corrosion, easy and efficient jointing by a variety of methods, ready availability in almost any shape or form, and high scrap value. Copper of the highest purity has been a more efficient conductor of current than any other known material except silver. Annealed copper that has a resistance equal to that specified by the International Electrotechnical Commission has been said to have a conductivity of 100% i.a.c.s. (international annealed copper standard), although in practice high-conductivity copper, which has been both wrought and annealed, can in fact be having a conductivity of 101% even 102% i.a.c.s. The resistivity of copper has been $1.56 \times 10^{-6}\Omega$cm at 0°C.

Copper Alloys : Mixtures of copper having small quantities of added elements. The impurities which may be found in high-conductivity copper could be controlled to within very close limits. Phosphorus in particular is a drastic effect upon conductivity, as little as 0.05% being sufficient for reducing the value to about 70% i.a.c.s.

Copper Loss : A term which is loosely applied to the conduction, joule or PR loss that occurs in conductors not intended specifically to produce heat.

Copper-oxide Rectifier : A metal rectifier has disks or plates of copper oxidised in a furnace at a temperature around 1.050°C and then annealed at a lower temperature. A layer of cupric oxide which gets formed on the outer surface is removed by controlled acid treatment, leaving a thin layer of cuprous oxide in contact with the copper matrix. It forms the semiconductor portion of the rectifier and good electric contact must be made over the whole of its surface.

Core :

(1) Refers to that part of an electromagnetic circuit which is situated within the winding.

(2) Refers to a single conductor with its insulation forming part of a cable.

Core Loss : Refers to the loss which gets developed as heat in the ferromagnetic core of a magnetic circuit subjected to alternating magnetisation. It consists of two components quite different in their origin, such as hysteresis loss and eddy-current loss.

Core Plate : Of a machine or transformer, this term refers to each of the thin, metallic sheets which is forming part of the core. The sheets are also termed as laminations, punchings, or stampings.

Core-type Transformer It is a transformer in which the windings surround and generally enclose the iron core.

Cored Carbon : A carbon rod which is used as an electrode in a carbon arc lamp. It is having longitudinal channels which is filled with mixtures to modify the effect of the arc.

Coreless : Applies to a device like an inductor or transformer without a core.

Coreless Induction Furnace : See high-frequency induction furnace.

Corona : Refers to a gas conduction process round a conductor which is caused by ionisation of the surrounding air. It provides in darkness, the visual appearance of a luminous sheath and appears when the potential gradient at the surface of the conductors gets exceeded a certain value. It causes a continuous rate of energy loss and may cause radio interference.

Corrosion : The term used for the chemical change of a metal from the elemental state into compound - either solids like oxides or sulphides, or salts like sulphates or chlorides in aqueous solution.

Coulomb : It is the SI unit of electric charge (symbol: C). IC–1As.

Coulomb's Law : This law may be stated that the force of attraction or repulsion between two electrically charged bodies has been proportional to the magnitude of their charges and inversely proportional to the square of the distance separating them.

Coulometer : It is an electrolytic cell which is also known as a volta-meter. It measures quantity of electric charge by the amount of a substance liberated as one of the products of electrolysis. It could be either a weight coulometer or a volume coulometer. Special types include a silver coulometer and gas coulometer.

Counter Compound-wound Motor : Alternating term for differentially coumpound-wound motor.

Counter-current Braking : It is a method of braking an electric motor by reversal of the supply connections. It is frequently termed as plugging.

Counter E.M.F. : Refers to the electromotive force which gets developed by certain forms of circuit element, by virtue of which they absorb electric energy and convert it into some other form such as mechanical or chemical energy.

Counter-E.M.F. Acceleration : Refers to a method of automatic acceleration which can be used when starting small d.c. motors. Use has been made of the fact that the counter e.m.f. across the armature of a d.c. motor increases in proportion to the speed as the motor accelerates. The operating coil of the accelerating contactor gets connected across the armature. When the supply gets connected at starting, the voltage across the contactor coil increases gradually until a value gets reached sufficient to close the contactor, and therefore short-circuit a step of starting resistance (see Figure 12). The duration of starting time has been proportional to the load on the motor.

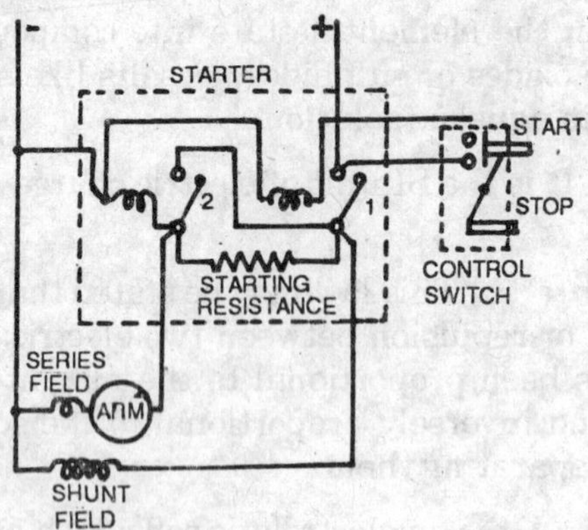

Fig. 12. Diagram of d.c. counter-e.m.f. starter.

Counterpoise Earthing : Refers to a method of earthing in which steel wires or tapes have been buried in the ground, the length and number depending upon the resistivity of the ground and the value of earth resistance required. This method finds use for protecting transmission and high-voltage distribution lines, and where ordinary rod or plate earthing has been unsuitable because of the nature of the soil.

Coupled Circuits : Refers to the circuits in which electric energy could be passed from one to the other. It implies that a simple two-mesh circuit gets coupled, because the two meshes are having one or more impedances in common. However, the term is generally restricted to circuits that have been separable, and particularly to those in which the coupling has been by means of a common magnetic field.

Coupled Surge :
Refers to a surge which is induced in a conductor by a surge in another adjacent conductor.

Coupling Coefficient : See coefficient of coupling.

Cradle : In overhead-line construction, this term refers to an assembly of guard wires to form a net.

Crater-lamp Oscillograph : An instrument which is used for recording the slow portion of a lightning discharge.

Creepage Distance : Refers to the shortest distance between two conductors along the surface of the insulating material that separates them.

Crest Factor Value : See peak factor, peak value.

Critical Damping : Refers to the least value of damping necessary to disallow oscillation in a system.

Crompton Potentiometer : It is a precise form of potentiometer. Resistance coils and switches have been replacing the long slide-wire of the basic potentiometer although in some a short slide-wire, commonly in circular form, has been retained as a fine adjustment. A simplified diagram is deficted in Figure 13.

Cross-compound Turbo-alternator : It is a turbo-alternator unit, which is generally restricted to large outputs. It consists of two lines, each with its own alternator. (cf. tandem compound turbo-alternator). The two lines may be having the same or dissimilar speeds.

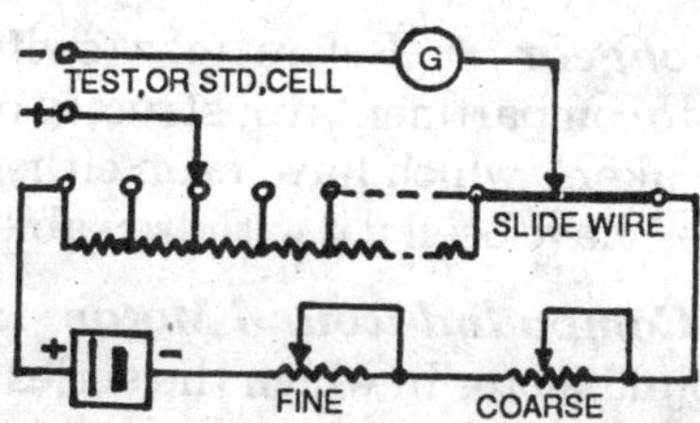

Fig. 13. Simplified diagram of Crompton potentiometer.

Cross-field Machine : It is a d.c. machine in which an essential feature has been the magnetic flux which gets produced on the armature axis, in addition to that produced on the field axis.

Cross-jet Pot : An explosion pot in which high pressures could be produced at a low current without exceeding a safe pressure at high currents. The gas formed in the early stage of contact movement produces making the oil flow across the path of the arc, which gets driven on to insulating barriers acting as arc splitters.

Cryogenics, Cryotechnique : The term used for the technique of operating devices and equipment at extremely low temperatures.

Cryotron : It is a superconductive device which consists of a fine super-conductive wire surrounded by a coil of fine wire having normal resistance characteristics. Superconductance in the first wire can get destroyed by the magnetic field produced by passing a current through a coil. The circuit also contains means for detecting whether, at a given instant, the wire has been superconducting or not. The complete device can be used in a computer as a binary memory element.

Crystal Diode : It is the combination of a crystal and a point contact for producing unilateral conductivity. Recent crystal diodes consists of germanium or silicon having a tungsten wire as point contact.

Crystal Lamp : Obsolete term for light-emitting diode.

Crystal Rectifier : Refers to a combination of a crystal and a point contact, or of two crystals, the junction having unilateral, or at least asymmetrical, conductivity.

Cubicle Switchgear : Refers to a switchgear which is divided into compartments constructed of steel sheet. The circuit-breakers, which have been either airbreak or oil-break types, have been not withdrawable as in truck gear.

Cumulative Compound-wound Motor : It is a compound-wound motor in which the series and shunt windings are assisting electro-magnetically.

Curie Point : Refers to that temperature, reached from cold, at which a given material ceases to be ferromegnetic. Such a material will regain its ferromagnetism on cooling, but the Curie point for falling temperature has been not always the same as that for rising temperature.

Current : Refers to a movement of electric charge. The direction of a current has been taken arbitrarily as that of the movement of positive charges and opposite to that of negative ones. A current is also resulted from the combined motion of positive and negative charges in opposite directions. In a conductor the current involves a drift of electrons towards the positive pole of the applied electric field. In an elecrolyte or in a gas it involves the migration of positive ions towards the negative electrode and of negative ions and/or electrons towards the positive electrode. The SI unit for current has been the ampere (symbol : A).

Current-control Acceleration, Starting : It is a method of automatically controlling the acceleration of resistance-started electric motors during the starting period. Each accelerating contractor is disallowed from closing until the starting current is falling to the minimum value.

Current Efficiency : In an electrochemical process, this term refers to the ratio of the mass of a substance affected chemically or electrolytically to the mass which would be expected from theoretical considerations.

Current Element : Refers to a short length of conductor carrying a current. It is often regarded in isolation, a concept which is fictitious but useful in that the total magnetic effect of a current in any conducting configuration can be obtained by summation of the effects of its constituent current elements taken separately.

Current-element Starter : See current-control acceleration, starting.

Current-Limiting Inductor : Refers to an inductor, in a power-supply system, which is arranged for series connection so as to limit short-circuit currents at various points on the network.

Current Pulsing : It is a method of producing large unidirectional, current pulses. A current-pulsing transformer is used. It resembles a transformer of normal construction, except that the secondary winding comprises a single turn of large section. With the secondary open-circuited, the primary gets connected to a d.c. supply. After the primary currents gets established, the secondary gets closed. The primary current has been then broken as rapidly as possible: a corresponding current gets developed momentarily in the secondary circuit.

Thus if a current of 10 A in a primary of 2000 turns gets suddenly broken, a current pulse appraoching 20,000 A would flow in the one-turn secondary. The underlying principle has been that of the constant-flux-linkage theorem. An essential part of a current-pulsing equipment has been a high-speed circuit-breaker which has been capable of interrupting corrents of the order of 10 A in about 2 ms.

Current-sharing Inductor : Refers to an inductor having a centre-tapped winding used in a static converter to get equal current distribution between two parallel operating converter devices.

Current Transformer : It is an instrument transformer which is used for the transformation of current. The primary winding gets connected in series with load to get measured or controlled, and this load decides the current flowing through it. The secondary winding has been loaded with a constant impedance, for any given set of conditions. The core flux and the current in the secondary circuit are dependent upon the primary current. It may be termed alternatively as a series transformer. Abbreviation: c.t.

Cut-out : It is a means of interrupting a circuit which is usually of low power. A fusible cut-out has been a fuse, but the term has been more usually applied to a magnetic device, such as that used in automobile electric systems to close and open the circuit between the generator and the battery in accordance with the effect of engine speed on the generator electromotive force.

Cut-wound Core : Refers to a magnetic-core construction which is using continuous grain oriented steel strip. The strip has been wound into a closed core of suitable shape, such as circular or rectangular, and impregnated.

Cyclic Rating : Refers to a form of cable rating. Tables of permissible current ratings (maximum permissible currents) for transmission and distribution cables have been based on the continued application of a steady alternating or direct current. In practice the maximum load has been usually applied only for a limited time, and the current may then get increased without the maximum permitted conductor temperature rise being exceeded. Such increases in current rating, where the applied load has been cyclic (usually with a period of once a day), have been given by the application of cyclic rating factors.

Cyclone Furnace : It is a boiler employing crushed coal. The ash get reduced to a fluid state so that it may get tapped and subsequently quenched to a solid state for disposal.

Cyclotron : It is an orbital accelerator which essentially of a vacuum chamber between the poles of a fixed field magnet. Inside this chamber have been two hollow Dee electrodes which load the end of a quarter-wave resonant line so that 10-20 MHz voltage appears across the accelerating gap. Protons or other positive ions get introduced from an ion source near the centre of the magnet and get accelerated twice per revolution as they spiral out from the centre of the machine.

Cylindrical Winding : Deprecated term for helical winding.

Damper :

1. A winding which consists of solid copper bars connected at the ends by partial or completed rings. It is inserted in the pole-face of a low-speed synchronous

machine. Its purpose has been to damp oscillations of angular speed about the mean (synchronous) speed. Under steady conditions the bars lie in, and rotate with, a constant magnetic field. If the speed has been subjected to cyclic disturbance, the bars cut the field and have e.m.f.s. induced in them. Currents would flow to oppose the speed variation by induction-motor torque action. It is also termed as an amortisseur winding.

2. A vibration-reducing device for overhead lines.

Damping : Refers to the oscillation decay of a freely oscillating system, which is caused by loss of initial energy of the system.

Dancing : The term used for overhead conductors when they oscillate due to the aerodynamic effect of ice accretions or of their profile. They may also be termed as galloping.

Daniell Cell : An early wet primary cell, which consists of zinc and copper electrodes in dilute sulphuric acid, with a porous pot having copper sulphate as a depolariser.

Daraf : Unit of elastance. It is reciprocal to capacitance. It is not practical use.

Dash Pot : Refers to a device which uses fluid friction to prevent sudden or oscillatory motion of a moving part.

D.C. : See direct current.

D.C. Amplifier : Refers to an electronic amplifier which has been arranged to deal with d.c. signals, or a.c. signals of very low frequency.

D.C.Balancer : Refers to a d.c. motor-generator or rotary transormer which is used in a multiple-wire sysem for distributing the voltage equally between the wires.

D.C. Bridge : Refers to a bridge circuit which uses direct current for the measurement of circuit parameters. The basic form has been the Wheatstone bridge.

D.C. Current Transformer : Refers to an arrangement which uses a transductor to provide a d.c. output proportional to the d.c. input.

D.C. Injection Braking : Refers to a method of braking an electric motor by disconnecting it from the supply and feeding direct current to the stator winding. The effect has been to build up a static magnetic field in he rotor space, and the current thereby produced in the rotor winding gives rise to a powerful braking torque.

D.C. Motor : It is achine which produces mechanical power when it is supplied with a direct current. There have been three principal types : the series-wound motor, the shunt-wound motor, and the compound-wound motor.

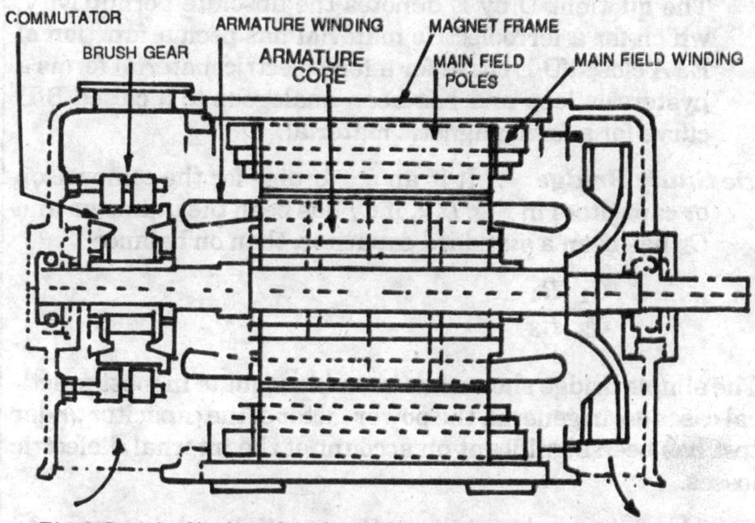

Fig. 1. Longitudinal section showing main components of a d.c. motor.

The field and armature-windings have been connected in parallel, with respect of the supply, for the shunt motor and in series for the series motor. Where a wide speed range with variable control has been needed d.c. motors would particularly suitable, as their inherent characteristics can be easily tailored to suit the application, and the control gear has been simple. A typical field of application has been electric transport vehicles which need a speed range from zero to maximum speed.

D.C. Resistance : The resistance offered by a conductor, or conductive device, to a constant direct current. The term has been employed in contradistinction to equivalent resistance, or a.c. resistance, in which either:

(a) the current/voltage characteristic has been such that the resistance r_d=v/i for a given voltage and current differs from r_a=Δv/Δi for small variations Δv and Δi, or

(b) the I^2R loss is increased, *e.g.* by eddy-current effects, which do not occcur with constant direct currents.

D.C. Voltage Transformer : It is an arrangement which is similar to a d.c. current transformer but with appropriate resistors in the input and output circuits.

D/E Curve : It is the graph of the relation between the electric flux density, D, and the electric field strength, E. The quotient D by E denotes the absolute permittivity, which for a ferroelectric material has been a function of E. A closed D/E curve for a ferroelectric material forms a hysteresis loop and has been analogous to a closed B/H curve for a ferromagnetic material.

De Sauty Bridge : It is an a.c. brdige for the comparison or capacitors in Fig. *D*.2, if C_1 has been the unknown and C_2 has been a standard capacitor, then on balance

$$\frac{C_1}{C_2}=\frac{R_4}{R_3}$$

The simple bridge shown has been inadequate in most practical cases as, in general, the power factor of the capacitor under test has been significant on account of the internal dielectric losses.

Dead : At or almost at earth potential and disconnected from any live system.

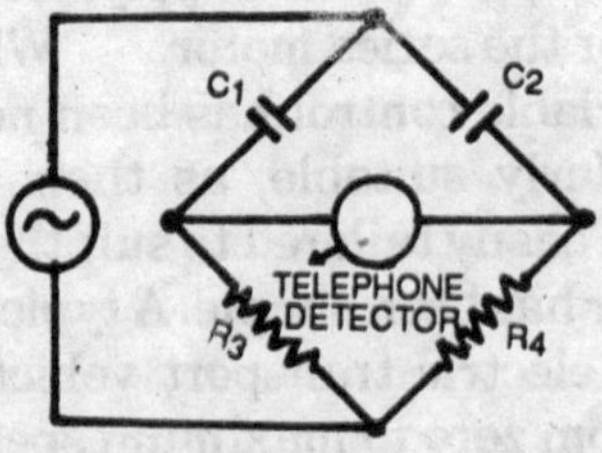

Fig. 2. De Sauty bridge

Dead Beat Mechanism : Refers to one in which the oscillatroy motion of its moving parts dies rapidly due to the damping present.

Dead Earth : Refers to a connection between a conductor and earth by means of a low-resistance path.

Dead-ended Feeder : Another name for independent feeder.

Dead-front Panel : Refers to a switchboard in which all the live parts of the switches, fuses, etc., have been mounted behind the panel.

Dead-man's Handle, Pedal : In electric traction, it is an attachment to the handle of a controller or to a brake valve, which has been normally held by the driver's hand or pressed by his foot. If the pressure of the hand or foot gets released, the device would act to cut off the current and apply the brakes. A more comprehensive safety device being fitted in locomotives has been the driver's safety device (DSD).

Decibel : One-tenth of a bel. It has been equal to ten times the common logarithm of a power ratio. It is a method which is used for expressing gain or loss in signal level in telecommunication circuits and acoustics.

Decomposition Voltage : For an electrolyte, this term refers to this minimum voltage between two electrodes immersed in it which will produce continuous electrolysis.

Decrement : Refers to a measure of the damping in a freely oscillating system where the decay has been regarded to be exponential. It is the ratio of two successively peak values of oscillation of the same sign.

Deep-bar Winding : Refers to a winding, especially a squirrel-cage widding on the rotor of an induction motor which has been designed to increase production of eddy-current loss and so increase the effective resistance. The additional loss which has been proportional approximately to the fourth power of the conductor depth, and a significant increase in starting torque would be obtained without much effect on the slip and efficiency at normal load. The power factor at full load, has been however, somewhat reduced due to the additional rotor slot reactance to which the generation of eddy currents is due Fig. D. 3 shows typical conductor shapes.

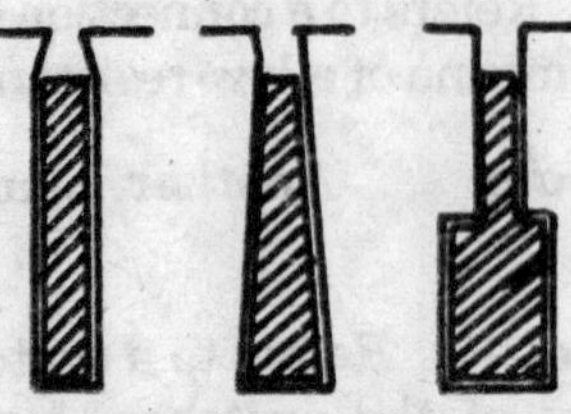

Fig. 3. Deep- bar conductors for a squirrel-cage winding

Deflection Potentiometer : Refers to a variation of the simple d.c. potentiometer in which a small deviation from a normal voltage would be read by balancing the potentiometer in the ordinary way for the nominal voltage, but employing a calibrated galvano-meter so as to read the actual departure from the nominal. The accuracy has been somewhat less than that of an ordinary potentiometer but better than that of a voltmeter.

Degaussing : Refers to the process of arranging current-carrying coils in a ship for counteracting its effect on the Earth's field, so as to avoid setting of the detector mechanism of maganetic mine.

Detion Grid : It is an arc-control device which finds use in certain air and oil circuit breakers. The contacts separate within a stack of U-shaped insulating plates. Some of the insulating plates are having an iron insert so that the arc current sets up a considerable magnetic field which encircles the arc and forces it towards the closed end of the U. This would cool the arc and breaks it up into a series of shorter arcs, facilitating its rapid extinction.

Delay : See phase-shift distortion.

Delay Angle : This term finds use with grid control of mercury-arc rectifiers. Transfer of the arc from anode in such a rectifier normally takes place at the instant in each cycle when the voltage of an idle anode starts to exceed that of the anode carrying the current. The transfer takes place quite naturally at such an instant, but by means of grid control it can get delayed by a fraction of a cycle, normally expressed in angular form as the delay angle. By delaying the transfer in this way *i.e.*, increasing the delay angle from zero, the mean value of the rectifier output voltage would get reduced. The same effect has been applicable to the use of triode thyristors.

Delay Cable : It is a concentric cable which gets connected to an object under test to delay the arrival of a surge from a surge generator.

Delay Line : Refers to any passive network which has been capable of delaying a signal of any waveform without introducing appreciable distortion, the parameters being selected to decrease the propagational velocity.

Delta Connection : It is a three-phase connection in which the three windings have been connected in series, the supply being applied to, or taken from, the junctions. It has been a special case of mesh connection, and has been sometimes written as a Δ connection.

Delta Voltage

1. Refers to the voltage between any two lines of a symmetrical three-phase system (also known as line voltage, or mesh voltage.
2. Also, refers to the voltage between alternate lines of a symmetrical six-phase system.

Demagnetisation Curve : Refers to the characteristic that ascertains the quality of a meterial for permanent-magnet purposes.

Demodulation : It is the inverse process to modulation. It is erroneously also called detection. It is involving the recovery of the modulating signal (or intelligence) from the modulated carrier wave after passage through the transmission medium.

Detection : Refers to the process of extracting information which is imparted on a carrier.

Diac. : The term used for a five-layer two-terminal gateless semi-conductor device as depicted in Fig. 4. If the applied voltage has been high enough, the device will start to conduct as a normal untriggered thyristor through layers $p_2n_2p_1n_1$. When the polarity of the supply gets reversed, a similar situation would occur but this time conduction would occur via layers $p_1n_2p_2n_3$.

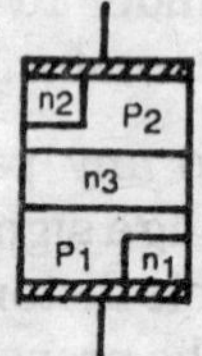

Fig. 4. Basic Diac.

Diamagnetism : A phenomenon which is exhibited by materials having a relative permeability less than unit.

Diametral Connection : It is a form of connection to brushes on a commutatos, or to tappings on a closed double-layer winding. The circuit has been taken through brushes or tappings which on a two-pole machine, have been diametrically opposite to each other; or on a multi-polar machine have been displaced from each other by 180 electrical degrees.

Diametral Voltage : Refers to the voltage between opposite lines of a symmetrical six-phase system.

Diaphragm :

1. Refers to a sheet of finely perforated or porous material which is used in a cell of an accumulator as a separator between the plates of opposite polarity.
2. A partition which is used in electrochemistry. It is permeable to ions.

Diathermic Coagulation : Refers to the use of current at high frequencies in electromedicine for destroying parts of the human body by heating to the point where albumen coagulates.

Diathermy : Electromedical treatment using sustained high-frequency current. A heating effect would be obtained throughout the tissues.

Dielectric : Refers to an insulant in which an electric field makes polarisation.

Dielectric Absorption : Refers to the property of a dielectric material due to which a charging or discharging current remains present, at an appreciable level, for a period longer than would be expected if the decay had been exponential.

Dielectric Breakdown : Deprecated term for electric breakdown.

Dielectric Constant : Obsolete term for relative permitivity.

Dielectric Heating : It is a method of high-frequency heating which is applied to insulating materials. When an alternating voltage has been applied across the plates of a perfect capacitor, an alternating current would flow which would lead the applied voltage by 90°, the power factor would be zero, and no power would be lost. A practical capacitor is having a dielectric material with losses, which appear as a shunt-resistance. A component of the current now flows in phase with the applied voltage,

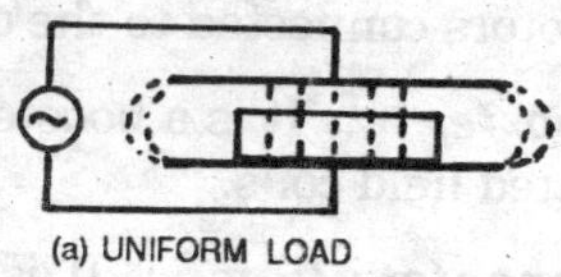

(a) UNIFORM LOAD

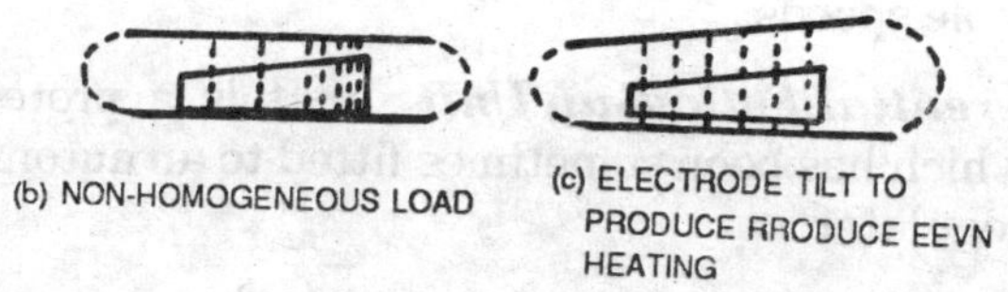

(b) NON-HOMOGENEOUS LOAD (c) ELECTRODE TILT TO PRODUCE RRODUCE EEVN HEATING

Fig. 5 Dielectric heating

and causes a power loss, which appears as heat throughout the dielectric. Fig. 5 (a) depicts the simplest form of dielectric heating, in which the workpiece to be heated is kept between two flat sheets of conductive material across which the high-frequency voltage is applied, and between which exists an electric field of uniform voltage gradient. If the non-conductive material forming the dielectric has been itself homogeneous, it will get heated up uniformly throughout. Fig. 5 (b) shows that eneven work kept between parallel electrodes will get heated non-uniformly and (c) depicts how a graded air-grip can overcome this trouble.

Dielectric Hysteresis : Deprecated term for electric hysteresis.

Dielectric Loss : Refers to the dissipation within a dielectric mterial when it has been subjected to a time-varying electric field.

Dielectric Polarisation : Deprecated term for electric polarisation.

Dielectric Strength Deprecated term for electric strength.

Diesel-electric Plant : Equipment using a generator which is driven by a diesel engine, *i.e.* a compression-ignition engine in which fuel oil gets injected into a heated compressed-air charge. In the case of a diesel-electric locomotive, the generator is able to supply direct current to electric motors connected to the driving axles.

Differential Booster : It is a booster which has differentially connected field coils.

Differential Concatenation : It is a method of cascade connection or concatenation, in which two motors may get adjusted to oppose one another, thereby giving four possible speeds.

Differential Follow-up Unit : It is a protective device which has been sometimes fitted to an automatic voltage regulator.

Differential Protection : It has been one of the most common and effective means which is used for the protection of a.c. systems and generators. It acts by comparing the current at the two ends of a feeder, or a machine or transformer winding, so that any divergence gets detected and relay action initiated.

Differential Selsyn : It is a unit which gets interposed in the electric tie between a conventional pair of selsyn units without direct connection to the supply. The differential selsyn is able to record angular displacement or rotates at a speed which has been either the sum or difference of that of the other machines. Conversely it has been possible to make the shafts of the transmitter and receiver selsyns rotate with a predetermined speed difference by driving the differntial machine at that difference speed.

Differential Windings : Refers to two windings on a piece of apparatus, so arranged that, when excited by direct current, their elecrtromagnetic effects have been opposed. Such an apparatus is termed as differentially wound or differntial.

Differentially Compound-wound Motor : It is a compound-wound motor in which the series and shunt windings have been opposed electromagnetically.

Diffused-junction Transistor : Refers to a transistor in which the width of the base has been very small, making the device suitable for high-frequency operation. It could be fabricated by the diffusion of of impurities into a solid crystal.

Dimensional Quantities : These are shown in Table 1.

Diode : It is a device with two electrodes, in particular an anode and a cathode, and a non-linear current/voltage characteristic which has been unsymmetrical.

Dipole : Refers to an arrangement of two separated quantities (*e.g.* two electric charges) of opposite polarity. If point charges +q and -q get separated by a small distance 1, such a system would be an electric dipole of dipole moment equal to q1. The term dipole has been applied to double-ended aerials, which in their simplest form have been a whole number of half-wavelengths long.

Direct-arc Furnace : It is an arc furnace in which the arc gets formed between three equally spaced graphite electrodes and the metallic charge. Compensating inductros have been kept in the circuit to take care of surges on the supply which is caused by solid metal coming into contact with the electrodes. Direct-arc furnace find use as steel-making units, smelting furnaces and copper-melting furnaces.

TABLE 1

Dimensions of electrical quaantities

Quantity	*Defining equation*	*Dimensional formula*
Energy	w	L^2MT^{-2}
Power	p=dw/dt	L^2MT^{-3}
Current	i	I
Current density	j=di/da	$L^{-2}I$
Voltage	v=p/i	$L^2MT^{-3}I^{-1}$
Electric field strength	E=-dv/ds	$LMT^{-3}I^{-1}$
Resistance	R=v/i	$L^2MT^{-3}I^{-2}$
Resistivity	e=R(a/s)	$L^3MT^{-3}I^{-2}$
Magnetic flux	$\Phi= -\int v.dt$	$L^2MT^{-2}I^{-1}$
Magnetic flux density	B=dΦ/da	$MT^{-2}I^{-1}$
Magnetomotive force	F=Ni	I
Magnetic field strength	H=dF/ds	$L^{-1}I$
Permeance	A=Φ/F	$L^2MT^{-2}I^{-2}$
Inductance	L=NΦ/i	$L^2MT^{-2}I^{-2}$
Permeability (abs.)	μ=A(s/a)	$LMT^{-2}I^{-2}$
Charge, electric flux	$q=\int i.dt$	TI
Electric flux density	D=dq/qa	$L^{-2}TI$
Capacitance	C=q/v	$L^{-2}M^{-1}T^4I^2$
Permittivity (abs.)	ε=C(s/a)	$L^{-3}M^{-1}T^4I^2$

Direct Cooling : Implies the application of a coolant direct to the conducting metal of a winding in a machine or trnsformer, by using hollow conductors.

Direct Copuling : Refers to coupling of coupled circuits in which the two circuits have been sharing a common impedance, or in which one circuit has been fed from the other through an impedance link.

Direct Current : A current which is having continuously the same direction of flow; in particular a current that is having a rigidly constant magnitude and direction. A unidirectional current of magnitude varying with time may be termed as a pulsating current if its variations have been rhythmic.

Direct Axis : The term used for the axis of the main field system in an electromagnetic machine which is electrically at right-angles to the cross or quadrature axis.

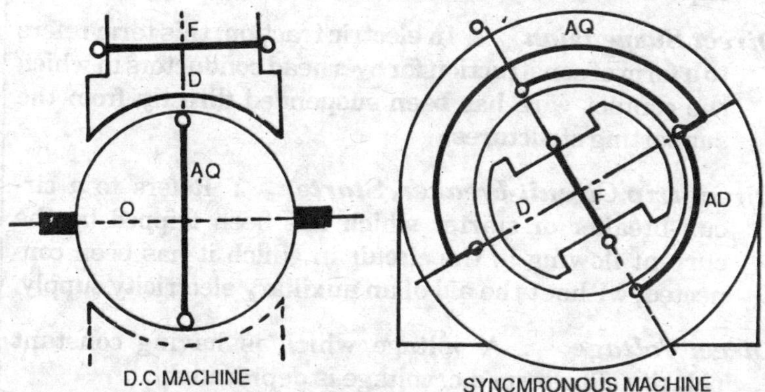

A armature F field D direct axis Q quadrature axis
Figure 6. Direct and quadrature axis

Direct Load Loss : The I^2R loss in an alternator stator which is based on the measured ohmic resistance corrected to 75°C.

Direct-on-line Starting : Refers to direct connection of a motor to the supply for starting. With single-phase motors it has been possible only in the case of self-starting machines like the series or repulsion types of commutator motor; with two and three-phase squirrel-cage motors it gets limited to small machines. Direct-on-line contactor

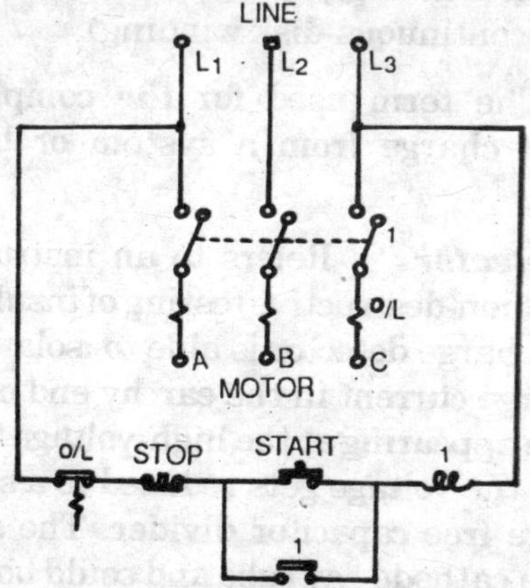

Fig. 7. Basic circuit of direct-on-line contactor starter for squirrel-cage motor

starters have been designed round the basic circuit of Figure 7. A hand-operated oil switch with under-voltage trip coil may be used with larger motors.

Direct Suspension : In electric traction, this term refers to a form of construction for overhead conductors in which the contact wire has been suspended directly from the supporting structures.

Direct-trip Circuit-breaker, Starter : Refers to a circuit-breaker or starter which has been tripped by the current flowing in the circuit in which it has been connected, without the aid of an auxiliary electricity supply.

Direct Voltage : A voltage which is having constant polarity. The term d.c. voltage is deprecated.

Disk Insulator : It is an insulator, in an overhead-line system, which has a number of separate units cemented to metal fittings which interlock to form a flexible string. It could be used as a suspension insulator or a tension insulator.

Disk Winding : Refers to a transformer winding which is used mainly for the high-voltage windings of medium and large transformers, in which the turns have been made up into a number of annular disks which have been wound singly and then joined (single-disk winding), or in pairs (double-disk winding), or from a continuous length of conductor (continuous-disk winding).

Discharge : The term used for the complete or partial removal of charge from a system or body or an accumulator.

Discharge Detector : Refers to an instrument which is used in the non-destructive testing of insulating material. An a.c. discharge detector is able to isolate and amplifies the discharge current in the earthy end of a specimen or the voltage appearing at the high-voltage terminal. In the latter case the voltage gets reduced to a suitable level by a discharge free capacitor divider. The signal gets displayed on a cathode-ray tube and could be compared with a calibrating pulse to get its magnitude and hence the discharge energy.

Discharge Lamp : Deprecated term for gas-conduction lamp. It is a source of light for illumination or for luminous signs. The colour of the emitted light has been characteristic of the gas content of the lamp; it has been yellow with sodium vapour, pale blue with mercury vapour, dark blue with neon and red with argon. Two factors affecting light output have been the current density and the gas pressure, both affecting the number of electron-atom collisions. With increase of gas pressure, increased voltage is needed to operate the lamp and the spectral emission also changes. To increase the flow of electrons, heated electrodes get introduced in all except the cold-cathode types, which have been run at higher voltage as a substitude for heating; all hot-cathode filaments are having chemical treatment to increase electron emission still more. Mercury and sodium-vapour lamps run hot and rely on the power in the lamp to produce sufficient heat to vaporise the metal. Mercury lamps operate off mains voltages of 200-260 V, but sodium lamps need a step-up transformer.

Discharge Tube : Deprecated term for gas-conduction tube. It is a tube of insulating material, usually glass, having two electrodes and evacuated to a sufficiently low gas pressure. It allows to pass a current when the voltage between the electrodes gets raised to a sufficiently high value.

Discharge Tube Rectifier : Deprecated term for gas-filled or vapour filled diode.

Discharger : Alternative name for spark gap.

Discrimination : The term used for the means by which protective equipment separate faulty apparatus from sound (which may, nevertheless, be carrying the fault current). Discrimination tripping signals have been sent only to those circuit-breakers whose operation would disconnect the fault from the system. Two basic systems of discrimination have been available graded-time delays, and the unit system.

Dispersion : Means to variation of capacitance with respect to frequency or time. In a dispersion test of an insulating material a range of 0.5-50 Hz for a sinusoidal

supply or 3-300 ms for a square-wave supply has been found to be suitable.

Displacement Current : Refers to a rate of change of electric flux, having a magnetic effect equal to a corresponding conduction current. Displacement current has been present in any material, conducting or insulating, whenever there has been applied electric field which changes with time.

Displacement Factor : Alternative term for power factor of the fundamental in a circuit which is having non-sinusoidal voltage and current. It is the ratio of the active power to apparent power of the fundamental components of voltage and current.

Disruptive Breakdown : In an insulating material this term refers to an electric breakdown which locally permanently destroys the insulating property of the material.

Disruptive Electric Field Strength : Of an insulant or insulating material, this term refers to the lowest value of electric field strength that makes disruptive breakdown.

Dissipation : Refers to conversion to heat and thereby loss of useful electric energy in a component, device or equipment.

Distorted Waveform : Generally, any waveform which has been not sinusoidal.

Distortion : The term used for the change in waveform that takes place when the scaled output response of an electric (or mechanical) signal transfer-system has been an imperfect reproduction of the input stimulus; for example, the change that takes place between input and output in an amplifier or transmission network.

Distortion Factor : Alternative term for harmonic factor.

Distributed Circuit Element : It is a circuit element whose characteristic parameters have been spatially distributed such as in a transmission line.

Distributed Winding : Refers to winding that is arranged uniformly over the surface of a stator or rotor, each coil possessing the same dimensions. In Fig. 8, a slot has one side of a coil, the other side being about one pole-pitch away, the actual being the span has been exactly a pole-pitch it has been a full-pitch coil, but commonly it has been less than this by one, two three or more slot pitches, being then known as short-pitched or chorded.

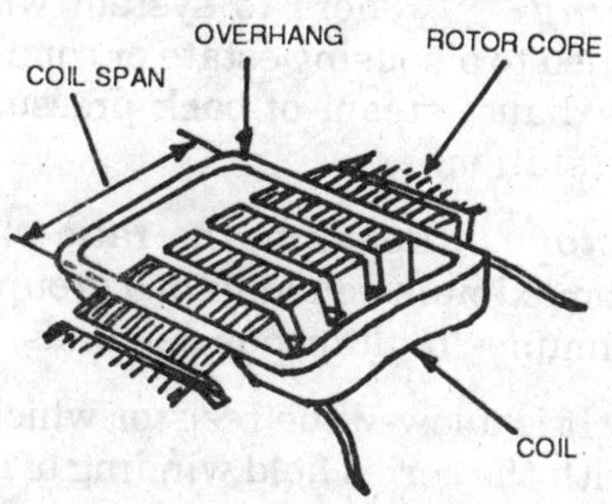

Fig. 8. Coil of a distrubuted winding

Distribution Board : Refers to an equipment which consists of busbars and possibly switches, fuses, links, etc., for connecting controlling or protecting a number of branch circuits fed from one main circuit of a wiring installation.

Distribution Factor : Refers to a factor which finds use in the calculation of the electromotive force generated in the winding of an a.c. machine. It use becomes necessary as the total e.m.f. has been less than the arithmetic sum of the e.m.f.s. in each coil, these not being in phase with one another.

Distribution Fuse Board : Refers to a distribution which has fuses in each of the branch circuits.

Distribution Pillar : A pillar which is having switches, links or fuses for interconnecting distributing mains.

Distribution Switchboard : A distribution board which is having switches in each of the branch circuits. These circuits usually also include fuses, and a main incoming switch may get incorporated in the board.

Distributor :

(1) A device which is used in automobile electric engineering to ensure that the voltage has been applied to the sparking plugs of an internal combustion engine in the correct sequence.

(2) Refers to the part of power distribution system to which consumers'circuits have been connected. It is also termed as a distributing main.

District Heating : Refers to system whereby heat has been supplied to a housing estate or commercial premises from the exhaust steam of back pressure turbines in a generating station.

Diversity Factor : Refers to the ratio of the sum of the individual maximum demands of a group of consumers to their maximum simultaneous demands.

Diverter : It is a low-value resistor which is connected in parallel with the series field winding of a d.c. machine so as to divert some of the current from it, either for control purposes or to correct some fault in design.

Dividing Box : Refers to a closed box in which the cores of multicore cable can get connected to external conductors.

Doctor : In electroplating, this term is used for an anode covered with a permeable material saturated with the plating solution. It has been applicable to parts of the object to be plated, the latter being made the cathode.

Dolly : Of a tumbler switch it is the operating switch which consists of a lever pivoted on the face of the switch It normally projects through an outer cover.

Donor Impurity : Refers to an element which is introduced into a semiconductor material having a valency than the semiconductor. It gives rise to free electrons which become available as negative charge carriers. This results in n-type conductivity.

Dose Meter : It is an instrument which is used for measuring quantity of radiation, particularly of X-rays in X-ray therapy.

Double Amplitude : Deprecated term for peak-to-valley value.

Double-break Circuit-breaker Switch : It is a circuit-breaker or switch in which there have been two breaks in series in each pole or phase.

Double Bridge : See Kelvin double bridge.

Double-cage Winding : It is a form of squirrel-cage winding which uses two (generally separate) cages, one set near the surface of the rotor slots and of high resistance, the other deeper and of low resistance. As the depth of slot is able to control the cage reactance, the outer cage is having high resistance with low reactance, while the inner cage is having the reverse. At starting the leakage reactance of the inner cage has been larger enough to cause the rotar current to flow chiefly in the outer cage, the resistance of which gives rise to considerable I^2R loss and therefore starting torque. If the speed has been normal the rectance of each cage would be negligible so that the rotor current gets shares between the two in accordance with their resistances. Typical rotor slottings are shown in Fig. 9.

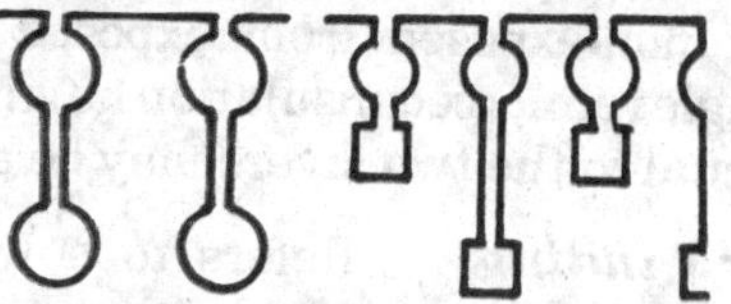

Fig. 9. Slotting of a double-cage winding.

Double Catenary Suspension : In electric traction, it refers to a form of construction for overhead conductors using two catenaries, supported at the same level from the same structure and possessing the same vertical sag. They form, with the contact wire, a triangular suspension formation.

Double-delta Connection : It is a six-phase connection in which the six windings could be represented diagrammatically as two triangles or two delta connections.

Double-element Wattmeter : It is a wattmeter which automatically is able to add two measurements of power, two moving systems being mounted on a common shaft. The method of connecting such an instrument for the meaurement of total power in a three-phase three-wire

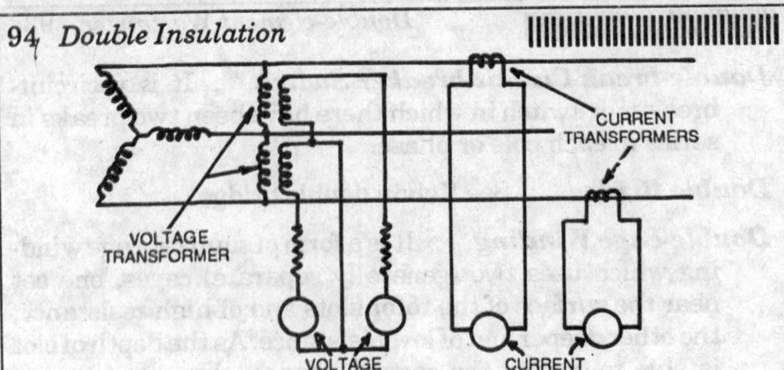

Fig. 10. Connections of a double-element wattmeter in a three-phase circuit

circuit having an unbalanced load, using voltage and current transformers, has been depicted in Fig. 10.

Double Insulation : Refers to appliance insulation in which external metal parts have been separated from live parts by two distinct layers of insulation-an inner or functional insulation which have been insulating live parts from each other and from non-exposed metal parts, and an outer or protective insulation which have been insulating non-exposed from exposed metal. In some cases a single reinforced insulation having characteristics at least equal to the two layers may be employed.

Double-layer Winding : Refers to a winding in which each slot is having two coil sides, one above the other.

Double-way Connection : In a static converter, it is a connection of the converter devices such that the current in the transformer secondary winding has been passing in both directions.

Down Conductor : Of a lightning protective system, this term refers to the conductor which is connecting the air terminations with the eath terminations.

Drag-cup Machine : It is a small machine (motor, generator or tachometer) having a drag-cup rotor. It is having essentially a hollow copper cylinder which is not attached to any rotor iron. The stator core and winding, and the 'rotor' core, have been both fixed, while the drag cup has been free to rotate in the intervening air-gap.

Drained Cable : It is a mass-impregnated cable from which the free impregnating compound is drained at a temperature above the maximum working temperature of the cable.

Draw-in Box : Refers to a through which cable can get inserted into or withdrawn from a conduit, sometimes including facilities for cable joining.

Drift Transistor : Refers to a form of transistor in which a field has been produced in the base region by arranging a resistivity gradient, so that carries have been swept across at high speed, decreasing the transist time and increasing the maximum frequency.

Driving-point Function : Refers to a type of system function which is related to the input terminals of a network.

Driving-point Impedance : Refers to the ratio of the voltage applied across a pair of network terminals to the current entering (and leaving) the network at those terminals.

Drooping-characteristic Welding Set : Refers to a welding set in which the voltage droops automatically from the striking voltage to the arc voltage as the current gets increased.

Drum Starter, Controller : Refers to a starter or controller which is having its moving contact parts arranged on a cylindrical surface.

Drum Winding : Refers to a winding which consists of coils arranged either on the inside or on the outside of a cylindrical core, and situated either on the surface or in slots.

Dry Cell : A lead-acid cell in which the sulphuric acid electrolyte has been completely absorbed and held by porous electrodes and separator material. The term is also applicable to a dry primary cell.

Dry Gas-pressure Cable : Refers to an unimpregnated paper-insulated internal-pressure cable whose space inside the lead sheath gets raised to a high pressure by using an inert gas in contact with the insulant.

Drysdale Potentiometer : Refers to a polar type of a.c. potentiometer. It has been basically of the Kelvin-Varley slide type for magnitude comparison, employing first a direct current and a standard cell, then alternating current and an a.c. galvanometer; a precision milliammeter has been forming the transfer instrument from d.c. calibration to a.c. use. The a.c. source has been a phase-shifting transformer, which has been manually adjustable over 90°, so that a phase-angle measurement could be made by reading the phase-shifter angular position when the potentiometer is balanced.

Quality : Refers to a parallelism between the equations and theorems of physical systems (in particular of electric networks). The impedance of a series *RLC* circuit could be written

$$Z=R+j\omega L+1/j\omega C$$

and the admittance of a parallel connection of pure *RLC* branches has been $Y=G+j\omega C+1/j\omega L$

The two expressions are same what of the same form; they are, in fact, the duals of each other. Hence the dual of circuit elements connected in series has been a parallel connection of the duals of these elements.

Duddell Oscillograph : The term used for the earliest type of oscillograph; it was also used as an oscilloscope. It is a mechanical device which is based on the combination of one or more galvanometer-type instruments having a rotating drum for photographic recording and a rotating mirror for visual observation. Now it is superseded by instruments using cathode-ray tubes.

Dummy Load : Refers to a circuit element which has been designed to simulate the usual load in a given electric circuit.

Duplex Winding : A winding consisting of two distinct windings which are connected to alternate segments of the commutator and joined in parallel by the brushes, as shown in Fig. 11 for lap-connected arrangement. The number of parallel circuits through a duplex winding would have been thus twice that for an ordinary simplex

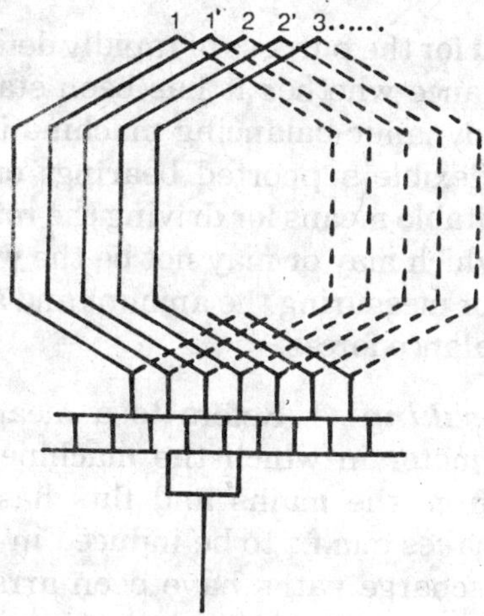

Fig. 11. Duplex lap winding.

winding. Duplex windings find use to increase the number of parallel paths through the commutator winding for high-current low-voltage machines.

Duralumin : Trade name for a high-tensile aluminium alloy having copper, manganese, magnesium and silicon. It has been used for the moving parts of switchgear.

Duration : Of a wave-front this term refers to the total time taken by a surge voltage or current to rise to its peak value from zero. It has been nominally 5/4 times the time interval between points where the voltage or current has been 10% and 90% of the peak value.

Dust Precipitator : See electrostatic precipitator

Duty, Duty Cycle : Refers to the sequence of operating conditions of a component device or equipment.

Dynamic Balancing : It is a means of determining and correcting unbalance in rotating machinery. If the axial length has been significant in relation to the diameter, static balancing could not be sufficient. Although the centre of gravity may be able to coincide with the axis of rotation and thereby show a state of static balance, the inertia axis may be at an angle to the rotational axis, and a dynamic couple of forces will get produced when the rotor revolves. Dynamic balancing machines have been produced for the purpose of rapidly determining the state of unbalance whether it has been static or dynamic or both. A dynamic balancing machine in its essentials is having flexible supported bearings on which the rotor runs, suitable means for driving the rotor at a convenient speed (which may or may not be the working speed) and means for measuring the amount and angular position of the unbalance forces.

Dynamic Braking : Refers to a means of breaking an electric motor in which the machine has been disconnected from the mains and flux has been established which makes e.m.f.s to be induced in the rotating windings. Discharge paths have been arranged so that currents flow, resulting in the loss of energy and subsequent braking of the motor.

Dynamo : A direct current generator.

Dynamometer : Refers to a device which is used for measuring the mechanical output of a prime mover or electric motor. An absorption dynamometer would absorb the output as well as measuring it, while a transmission dynamometer is able to measure the output transmitted to some other equipment. A dynamometer is also an electric instrument that is able to measure the force exerted between fixed and moving coils.

Dynamometer Relay : Refers to a moving-coil relay in which the field flux has been produced by one control current, the second control current being applied to the moving coil which operates the contacts. The operating force has been proportional to the product of the fluxes, and by using spring control the contact movement has been made proportional to the product of the two input currents. If the two coils have been carrying the same current a square-law response is obtained.

Dynamometer Wattmeter : It is an instrument which is used for precision and industrial measurement of power at power frequencies. As a precision instrument it could be constructed so that it is equally accurate when used with direct or alternating current and voltage.

Dynamotor : Alternative name for rotary transformer.

Ear : In overhead construction, the term refers to a metal fitting that has been attached to a contact wire to suspend the wire or hold it in position. Various type have been the straight-line ear, anchor ear, splicing ear, etc.

Earth : Refers to the conductive mass of the Earth or any conducted connected to it. The main body of the Earth has been regarded to be a zero potential.

Earth Current :

1. Refers to a fault current which flows to earth.
2. Refers to a large current which is circulating in the body of the Earth. Such currents may be induced by changes in the Earth's magnetic field. Variations of the magnetic field by up to 0.5×10^{-4} tesla for two or three days, due to magnetic storms related to sunspots, have been able to induce current of up to 5000 A.

Earth Detector : An instrument which is used for measuring or indicating a leakage current to earth. It is also termed as a leakage indicating.

Earth Electrode : Refers to a metal plate rod, strip etc. which is burried in the ground to make contact with the general mass of the Earth, usually for protective purposes.

Earth Fault : Refers to an accidental connection, or low resistance path between a live conductor and earth.

Earth-fault Protection : Refers to a system of protection whereby, should an insulation defect occur making earth-leakage current to flow, the faulty circuit has been automatically disconnected by using a fuse or circuit-breaker.

Earth-leakage Circuit breaker : It is a device which is able to disconnect the supply if the voltage on non current-carrying metal work or in the out-of-balance current in the supply due to leakage becomes greater than a predetermined value. Such a system could be operated more rapidly and at lower values of fault current than one depending on over-current protective devices. It does not need a low-resistance connection with earth.

Earth Return : Refers to the return path of an electric circuit which is having only one insulated conductor. The current returns to the source through the earth.

Earth Shield : Of a cable, the metal sheath which is immediately inside and in contact with the lead sheath.

Earth Termination : Refers to the part of a lightning-conductor system which has been designed to collect discharges from the general mass of the earth or to distribute charge into the earth. It comprises of all the parts below the testing joint in a down conductor.

Earth Testing : It is a process which involves (a) the measurement of the resistance to earth of earth electrodes, plates, rods, strips, etc. buried in the ground to make contact with the main body of the earth, usually for protective purposes; (b) the measurement of the resistivity of the soil itself, for use in the design of an earth-electrode system, and in investigating the nature of the underlaying strata. The resistance of any earth electrode has been made up of several components those of the connecting lead the contact between the electrode and the soil, and the body of earh surrounding the electrode.

Earthwire : Refers to a conductor which has been connected to earth and which runs parallel to and in close proximity with the conductors of an overhead transmission line.

Earthed System : Refers to a system of distribution in which the neutral point or one conductor has been maintained at earth potential.

Earthing : Refers to low-impedance connection of part of a system or equipment to earth. The main aim of earthing has been (a) to ensure the security of the system by limiting the potential of the live conductors, with respect to earth, to values that their insulation will withstand, (b) to safeguard life and property by ensuring that dangerous potentials with respect to earth are not maintained on parts of electric equipment or on non-electric metalwork normally not alive.

Earthing Inductor : Refers to an inductor which has been connected between the midpoint or neutral point of a system and earth for limiting the earth-fault current which may flow.

Earthing Switch : A device which used to connect to earth an isolated circuit to ensure safety of maintenance personnel. It is frequently combined with an isolator which is having a blade arranged to engage fixed earth contacts. On high-voltage outdoor isolators it is having a blade which is hinged to the base of the isolator to connect the live end of a post insulator to earth.

Earthing Transformer : A transformer which is intended primarily to provide a neutral point for earthing purposes.

E-class Insulation : Refers to one of seven classes of insulating materials for electric machinery and apparatus on the basis of thermal stability in service. It is particularly applicable to motors.

Eddy Current : Refers to a parasitic current which flows in a closed conducting path. It is generated by a magnetically induced e.m.f.

Eddy current Brake : Braking system which is mainly used mostly in energy meters or absorption dynamometers. In these cases the magnetic field has been developed by a permanent magnet or d.c. electromagnet, and the eddy currents are produced by the movement of a conductive mass across the field. For convenience the mass has been usually rotatable disk.

Eddy-current Heating : The form of heating which is involved in induction heating.

Eddy-current Loss : Refers to an I^2 R loss which gets resulted from current induced in a conductor. There is a loss in the iron of any magnetic circuit carrying an alternating flux.

Edison Screw Cap. : Refers to a lamp cap in which one contact has been the outer case, in the form of a screw thread, and the other contact has been a central projection.

Effective Resistance : Deprecated alternative term for equivalent resistance.

Effective Valve : Of an alternating current of voltage, it refers to a valve which is equal to its root mean square valve.

Effectively Earthed System : Refers to a system in which during a line-to earth fault the maximum voltage to earth of the other two lines does not become greater than 80% of the normal line-to-line voltage. This is the case when, for all switching conditions and unlimited supply the ratio of zero phase sequence reactance to the positive phase sequence reactance has been less than three, and the ratio of the zero-sequence resistance to the positive-sequence reactance has been less than unity. A system in which all the transformer are having star-connected windings with all neutrals directly earthed has been considered as being effectively earthed.

Efficiency : Of a device, equipment or plant where energy gets converted form to another,the ratio of the energy delivered at the output to the energy supplied at the input. The ratio is generally expressed in percentage of the input.

Elastance : Refers to the reciprocal of capacitance.The term has been not in practical use.

Elbow : Refers to a conduit fitting. It is a short radius length of tubing for connecting two lengths of conduit at right-angles

Electret : Refers to a the electric counterpart of a permanent magnet a piece of material which has a permanent bulk electric moment. It is generated by the solidification, in a strong electric field, of mixtures of certain organic waxes.

Electric Breakdown : Refers to the abruptly occurring change of an insulant from a non-conductive to a conducting medium when the insulant has been subjected to an electric field, of sufficient strength. The change could be irreversibily destructive.

Electric Charge : See charge.

Electric Constant : Refers to the quantity ε_0 which is the SI is having the approximate value $e_0 = 8.854$ pF/m. It is also known as absolute permittivity of vacuum and has been related to the magnetic constant μ_0 and the speed c_0 of light in vacuo by the expression

$$\varepsilon_0 \mu_0 c_0^2 = 1$$

Electric Fence : Refers to a fence which consists of a single wire, generally barbed, suspended on posts at a convenient height and no-where is direct contact with the ground, which carries a high frequency high-voltage charge from a small battery-operated charging set in a weatherproof box. If an animal touches the wire it would receive a shock but is not harmed. A screen to keep fish from powerstation water is consisting of a row of aluminium electrodes that discharge a mild current into the surrounding water.

Electric Field : Refers to the phenomenon associated with an electric charge. An electric charge would generate in the surrounding space, an energy state such that mechanical forces have been exerted on other charges. The charges have been said to be in an electric field, and the concept can be used for explaining electric phenomena without always referring them back to the charges by virtue of which they occur. By an extension of the concept, the electric field has been appearing as a constituent of electromagnetic radiation.

Electric Field Strength : See voltage gradient.

Electric Flux : Refers to the surface integral of the electric field strength, normal to the surface. It is regarded to be in a direction outwards from a positive charge.

Electric Furnace : It is a furnace which is supplied with electric energy and operating either without melting the charge (for example.)

Electric Hysteresis : Refers to the phenomenon which is observable in some dielectrics in which their electric state at an instant of variation has been related to their previous state. A symmetrical hysteresis loop for an electret material has been analogous to the symmetrical hysteresis loop for a permanent magnet material.

Electric Induction : Refers to the modification of the distribution of charge in a conductive body when it is subjected to the influence of an electric field.

Electric Machine : Refers to an electromachanical device termed as a generator when converting machanical to electric energy, and as a motor when converting electric to mechanical energy.

Electric Polarisation : Refers to the change of the physical state in a dielectric when subjected to an electric field. Each small element is becoming an electric dipole.

Electric Resistance Welded Tube : A tube which is manufactured by continuously butt seam welding two edges of steel strip. Linear speeds up to 50 m/min. have been in use, and the product has been some-times referred to as ERW tube.

Electric Sensor : Refers to a device that converts a non-electric signal of any kind to an electric signal.

Electric Shock : Refers to sensation which is produced on the nervous system by the passage of an electric current through the body. The severity of the shock is dependent primarily upon the magnitude of the current, the path which the current takes through the body, and the duration of the contact. In extreme cases the shock cause failure of the normal action of the heart and lungs, giving rise unconsciousness or death. Prolonged exposure to a current of 10—30m A will in most cases be lethal; a current of 100 mA can kill instantly if it passes through the heart. Shocks experienced from direct and alternating current have been in their effects, alternating current at normal frequencies (25—50 Hz) being the most dangerous. With the increasing use of equipment operating a high frequencies, an added danger would result passage of h.f. current through the body. At a frquency of about 100 kHz, the sensation ofshock starts disappearing, although serious internal burn can result.

Electric Space Constant : Term superseded by electric constant.

Electric Thread : Refers to a special form of thread which is used on screwed steel conduit.

Electrically Exposed : Term applied to an installation, generally one connected to an overhead line, in which the apparatus has been subject to over-voltages originating in the atmosphere.

Eletricity : The term is used to describe electric energy supplied as a commodity like fuel. It is also applicable to the particular field of science and technology which deals with electric phenomena.

Electroascoustical Transducer : A transducer which is designed to receive an input from an electric system and provide an output to an acoustical system, or vice versa. Well-known examples have been the microphone and loudspeaker.

Electrocardiograph : It is an instrument which amplifies and records the very small fluctuating voltages set up between various parts of the body because of the beating heart. It consists of a highgain electronic amplifier, to the output of which has been coupled a direct writing chart recorder. As the pulses have been of a low frequency the amplifier has been generally designed for a linear response from about 1 to 50 Hz, although some have a wider range from 0 to 100 Hz. The record produced has been termed as an electro cardiogram.

Electroceramic : A material which is made from powders of inorganic compounds which is chosen on the basic of their electric and magnetic properties.

Electrochemical Equivalent : Refers to the mass of a substance which is liberated by passage of one coulomb of electric charge.

Electrochemical Machining : Refers to the machining of metals by electrolytic action. The item to be machined has been made the anode in a fast-flowing electrolyte (eommonly caustic soda). By using a shaped cathode and a heavy current, metal has been removed at a high speed.

Electrochemical Series : Refers to a classification of the elements in order of their potential when immersed in a solution of normal ionic concentration.

Electrode : Refers to a conductor to or from which a current flows; for example, the anode and cathode of an electron tube or electrolytic bath.

Electrode Bar, Electrode Wheel : Of a resistance seam-welding machine, alternatively known as contact bar, contact wheel.

Electrode Boiler : A steel tank having water in which are kept three or more electrodes, the electrodes being connected to an a.c, supply. Current has been passed between the electrodes through the water and so the water gets heated.

Electrode Voltage : Refers to the voltage between an electrode of an electronic device and its cathode. Sometimes it is erroneously called electrode potential.

Electro-deposition : Refers to the deposition of a metal, alloy or compound by electrolysis.

Electrodialysis : It is an electrolytic method of detailing water. A direct current would pass through an ionised solution by means of ion migration. Charged semi-permeable membranes would separate positive sodium from negative chloride ions.

Electro-diesel : It is railway locomotive which incorporates a diesel engine and electric traction motors. It is capable of working either direct from an external electric power supply or with the diesel generator supplying the traction motors.

Electrodynamic Instrument : It is an indicating instrument which is similar to a moving-coil instrument in which the permanent magnet has been replaced by a fixed coil system (see Fig. 1). Both fixed and moving coils have been connected to the supply to be measured, rendering the instrument of use for low frequencies. In some fixed laboratory instruments thed range may be up to about 20

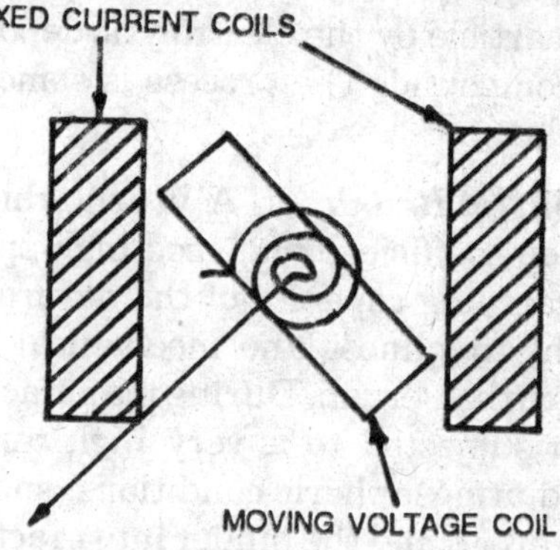

Fig. 1. Electrodynamic wattmeter

kHz. Pivoted instrument have been of Particular application as wattmeters, but as voltmeters and ammeters they are not having advantage over the moving-iron type of power frequencies, and they have been more expensive. Electrodynamic instruments may be having no core or an iron core. The latter is more recent, using nickel iron of low loss, and additional magnetic flux has been provided by the windings through the reduced reluctance, giving rise to a higher torque for a given consumption. The scale shape would be linear for a wattmeter but square-law for an ammeter and a voltmeter. True r.m.s. value has been indicated. The instrument has been very useful as an a.c./d.c. transfer instrument in the laboratory.

Electroencephalograph : An instrument which is used for recording, in the form of an elecrtroencephalogram, the electric activity of the brain. This activity could be detected in the form of tiny fluctuating voltages taking place between various points on the scalp. The instrument has a number of high-gain amplifiers each feeding a pen on a recording chart. This arrangement allows simultaneous recording from various parts of the brain. The frequency response of the amplifier would be normally from 0 to about 75 Hz.

Electroextraction : The term used for the direct recovery of a metal from a solution of its salts by electrolytic means. The process is also termed as electro-winning.

Electroforming : Refers to the production or reproduction of an article by the electrolytic deposition of a metel, alloy or compound. The process is sometimes termed as galvanoplasty.

Electrographitic Brush : A brush which combines the low friction coefficient and good high-speed running and current-carrying capacity of the natural graphite brush having the toughness and mechanical properties of the ordinary carbon brush. During manufacture the material has been subjected to a very high temperature under controlled atmoshpheric conditions, so transforming the original carbon and the binder into practically pure artificial graphite. The material has been produced with various degrees of graphitisation.

Electrohydrodynamic Generator : Refers to a generation device in which a stream of gas gets ionised, the positive ions being carried away by the stream while the electrons having collected by an electrode ring making a current to flow through a wire between the ring and a collecting grid kept downstream.

Electroluminescence : Refers to the emission of light from a phosphor under the direct action of an electric field. In practical devices the phosphor, in the form of a fine crystalline powder dispersed in a resinous insulant, gets sandwiched in a thin layer between two conductive surfaces to form a capacitor. One surface (or electrode) has

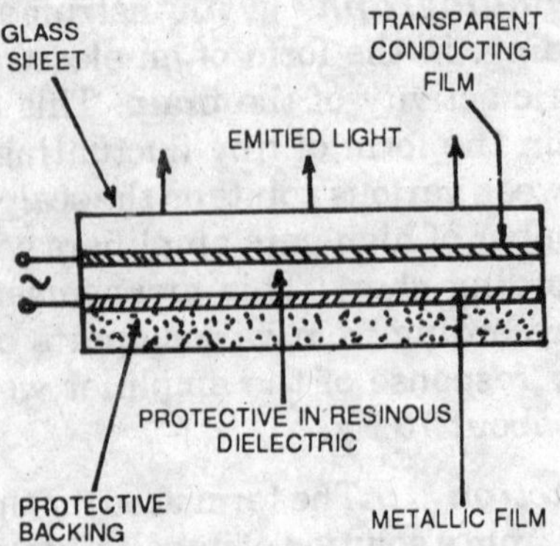

Fig. 2. Electroluminescent panel. For the purpose of clarity the vertical thickness in this diagram is exaggerated

been transparent, to allow the light from the phosphor to emerge. In one form of construction the supporting member has been a glass sheet, one surface of which has been suitably treated to form the transparent conductive electrode. The other essential layers have been built up on this substrate, as shown in Fig. 2.

Electrolysis : Refers to the decomposition of an elecrolyte by the passage through it of a direct current. It involves a transfer of ions between the two electrodes. The primary chemical reaction at the anode, or positive electrode may involves the dissolution of the metal of the anode or the evolution of oxygen or chlorine or some other gas. At the cathode, or negative electrode, the primary reaction may involve the evolution of hydrogen or, if the electrolyte is the solution of a metallic salt, the deposition of the metal component.

Electrolyte : Refers to a substance which conducts by virtue of its ion content. It is usually either a liquid or a substance which gets first dissolved in a suitable solvent, usually water.

Electrolytic Capacitor : Refers to a fixed capacitor in which the dielectric has been a thin film of oxide deposited upon aluminium foil which forms the positive plate, the effective negative plate being a non-corrosive electrolyte. The great advantage of this type of capacitor has been that a large capacitance could be obtained in a component of small dimensions.

Electrolytic Cell : A vessel having a system of electrodes and an electrolyte for the purposes of electrolysis.

Electrolytic Diaphragm : Refers to a partition which is used in electro-chemistry that allows the passage of ions but prevents conducting solutions from mixing freely.

Electrolytic Dissociation : Refers to the separation of certain substances into oppositely charged ions, or ionisation.

Electrolytic Meter : It integrating meter, the operation of which depends on electrolysis.

Electrolytic Rectifier : Refers to a type of rectifier which is now generally displaced by the metal rectifier. Its operation has been based on the property of unilateral conduction possessed by certain combinations of metallic or non-metallic electrodes. One form has been the aluminium rectifier.

Electrolytic Tank : Strictly, a bath having an electrolyte and electrodes, *e.g.* for electrochemical processes; more usually, a device for the plotting of electric conduction fields.

Electromagnet : Refers to a magnet which is excited by a current in a coil surrounding a ferromagnetic core, in contrast to a permanent magnet. The majority of practical magnets have been of the current-excited type, because the current provides a simple control of the amount and sense of the magnetic flux produced.

Electromagnetic Force : Refers to the mechanical force between currents, which is developed through the medium of a magnetic field. Mechanical forces have been developed in such a way that any resulting movement increases the flux-linkage with the electric circuit, or lowers the magnetomotive force necessary for a given flux. In the former case an increase of linkages needs more energy from the circuit, making mechanical work available in the latter, stored magnetic energy is released in mechanical form.

Electromagnetic Induction : Refers to the production of an induced voltage in a circuit by using a chnage in the magnetic flux through the circuit. An induced current can get produced in a closed circuit in a variety of ways

(a) by moving a magnet near the circuit,

(b) by moving the circuit near a magnet,

(c) by altering the size of the circuit,

(d) by varying the strength of the magnetic field. All these situations have been covered by Faraday's law of electromagnetic induction.

Electromagnetic Machine : Refers to a energy-converting device in which the principles of electromagnetic force and induction have been such devices have been including mechanisms of the relay type, electromagnetic pumps, loud-speakers, etc. as well as rotating machines.

Electromagnetic Pump : A device which is used for pumping liquid metals. The good electric conductivity of these makes it possible to maintain electromagnetic forces directly within the liquid itself, so avoiding the need for mechanical impellers. The field may be alternating or steady, and the current either supplied directly from an external source (see conduction pump) or induced by the field.

Electromagnetic Radiation : Refers to the wave propagation of energy in electromagnetic form.

Electromagnetic Screen : Refers to a component which is made of conductive material designed to surround an electromagnetic device and to function as a barrier against the penetration of a varying electromagnetic field in the direction from inside to outside the screen or in the reverse direction.

Electromagnetic Spectrum : The range of wavelengths of electromagnetic radiation.

Electromechanical Braking : Refers to a system in which brake shoes have been applied to the moving drum by using springs or weights, and released electrically.

Electrometallisation : Refers to the deposition, by electrolytic means, of a metal on a non-conductive base for decorative or protective purposes.

Electrometallurgy : Refers to the smelting, melting and refining of metals by the application of electric energy.

Electrometer : An instrument which is used for the measurement of electric charge. It has two electrodes capable of relative movement and connected to the source of charge to be measured. Normally one electrode has been fixed and the other has been suspended or otherwise mounted so as to get attracted or repelled under constraint. For maximum sensitivity only very limited movement has been allowed and this, commonly a rotation, may be read by a microscope, or with a lens, mirror, lamp and scale (Fig. 3).

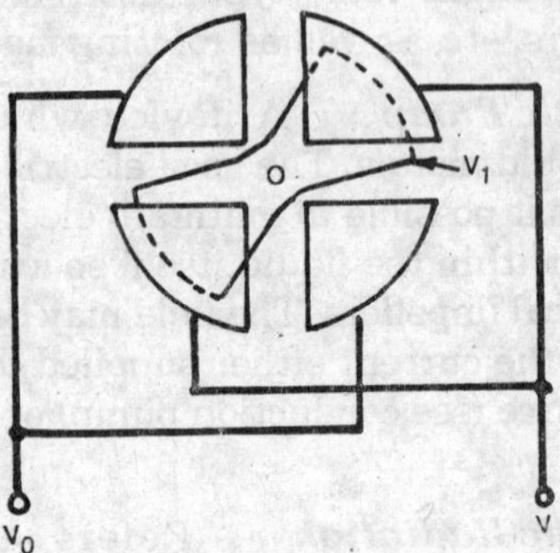

Fig. 3. Quadrant electrometer. V is unknown potential; V_0 is known low potential V_1 is known high potential.

Electrometer Valve : An amplifying valve having very high insulation resistance and impedance and very small grid current.

Electrometer Voltmeter : A form of valve voltmeter is using an electrometer valve.

Electromotive Force : The voltage E_0 of the source of the equivalent voltage generator depicted in Fig. 4. It makes the passage of current in a closed circuit and its direction is form the positive to negative terminal of the source. Its abbreviation is e.m.f. There has been a tendency to use the term source voltage instead of electromotive force, because the latter is a misnomer for voltage.

Electromyograph : An instrument which is used for amplifying and recording the electric signals associated with muscular action. These very small voltages have been detected either by means of a special needle inserted into the muscle or by two contacts spaced about 2 cm apart on the surface of the skin over the muscle.

Electron : It is a fundamental constiuent of matter which is having wave-particle dualism. In practice, it may be considered as

(a) an electrically charged particle,

(b) a mass point with no spatial extension, or

(c) an element of charge possessing intertia. Nuclear

bombardment experiments involving high-energy electrons indicate a major dimension of the order of 10^{-14}m. The charge is negative and of magnitude $e=1.602P\times10^{-19}$C. The rest mass has been $m_e=9.109\times10^{-31}$kg, and the charge-to-mass ratio is

$e/m_e=1.7588\times10^{11}$ C/kg

Electron-beam Furnace : It is a electric furnace of the melting type in which an accelerated stream of electrons in a vacuum get impinged on the material to be melted. It can be used to melt and refine metals having very high melting points, like niobium and tantalum.

Electron Cyclotron : Refers to an orbital accelerator in which electrons get accelerated in a vacuum chamber between the poles of a fixed field magnet. The orbits are having a series of discrete circular arcs which have a common tangent at a resonant cavity in which the electrons gain their successive increases of energy from a high-frequency electric field. Upper limits of energy achievable have been below 50 MeV with means currents below 1 μA.

Electron Diffraction : The diffraction of electrons which is obtained when a beam is passed through thin metal foil, the metal crystals forming a diffraction space lattice. The diffraction of electrons has been a natural outcome of their wave properties; the diffraction peaks have been arising from the reflection of electrons waves by planes of atoms in the crystal. Techniques similar to those used with X-rays have been used but since electrons react more strongly with matter than X-rays (owing to their charge), and have therefore less penetrating power, electron-diffraction methods have been confined to surface investigations such as contaminant films on metals.

Electron Gun : A structure having an electron-emitting cathode which in combination with an electron lens has been designed to produce a focused electron beam for use, for example, in a cathode-ray tube or in an election microscope.

Electron Lens : It is an electric or magnetic field distribution, or combination of the two, which is having an effect on an electron beam similar to that of an optical lens on a beam of light. An electrostatic lens has an electric field which is produced by two or more electrodes of suitable shape and dimensions and maintained at suitable potentials. A magnetostatatic lens is having a narrow current-carrying coil so shielded magnetically that the magnetostic equipotentials have been confined to a short axial distance within the coil.

Electron Microscope : Microscope using an electron beam instead of the more common light beam. High-velocity electrons may pass through atoms without undergoing deflection or other interference, and magnetic or electric fields may be used to render a beam of electrons convergent or divergent (see electron lens). The particular advantage of an electron microscope has been the enormous magnification of which it has been capable. The wave nature of light limits the resolution attainable in light ray microscopy to approximately 0.2 μm, but the electron microscope may attain a resolution of 10^{-3} μm.

Electron Mobility : Refers to the drift velocity of free electrons in a conductor per unit voltage gradient.

Electron Optics : Deals with the treatment of the paths of electron or ion beams in appropriate distributions of electric or magnetic fields, or combinations of both, in a manner analogous to the treatment of the paths of light rays in refractive media.

Electron-volt : May be defined as the energy acquired by an electron when it gets accelerated by a potential difference of one volt. 1 eV$=1.6\times10^{-19}$ J$=4.4\times10^{-26}$ kWh.

Electronegative : See anodic.

Electronegative Gas : It is a gas whose molecules have affinity for free electrons with which they combine to form negatively charged ions. This property makes it applicable for use as an arc-extinguishing agent in high-voltage circuit-breakers. An outstanding example has been sulphur hexafluoride.

Electronic Oscillator : It is a device which is used for producting a.c. power at some required frequency. It comprises as essential constituents an electronic vacuum tube (valve) or semiconductor device and a frequency-determining circuit.

Electronic Tube or Valve : A device having two or more electrodes within an envelope which has been either highly evacuated, hence vacuum tube (valve), or after evacuation gets filled with a low-pressure specific gas or mercury-vapour (generated in operation), thus gas or vapour-filled tube (valve). Vacuum conduction, by the passage of electrons only, occurs in a vacuum tube (valve), and gas conduction takes place in a gas or vapour-filled tube (valve).

Electronics : Refers to the study of the conduction of charges in a vacuum, in gases and in semiconductors, and the design and application of devices whose actions are consequent upon this feature. In fact, the boundary line between electronics and 'light' electric engineering has been hazy, to say the least.

Electroparting : Refers to the separation of metals by electrolytic means.

Electrophoresis : Alternative term for cataphoresis.

Electrophotoluminescence : An effect allied to electrolum- inescence. It includes varying degrees of enhancement or quenching by electric fields of both fluorscence and phosphorescence in zinc sulphide and other phosphors.

Electroplating : Refers to the deposition, by electrolytic means, of a metal on a metallic surface for decorative or protective purposes.

Electroprecipitation : Refers to the diposition of a colloidal substance by cataphoresis.

Electroscope : Instrument which is used for indicating an electric charge or potential difference by electrostatic means. A simple form is having two strips of gold or aluminium foil supported in a container by an insulated wire. When the wire get charged, the strips gets separated.

Electrosmosis : Refers to the passage of a liquid through a diaphragm under the stimulus of a difference of potential. It is also termed as electro-endosmosis.

Eelctrostatic generator : A generator which is based upon electrostatic action (see Van de Graaff generator, Wimshurst machine). It may also be termed as an influence machine or static machine.

Electrostatic Instrument : Indicating instrument that uses the electrostatic attraction between oppositely charged conductros or the repulsion between similarly charged conductors to measure relatively high voltages. The attraction type has been commonest, usually comprising an interleaving system of fixed inductors or quadrants and a moving system of vanes forming a variable capacitor; this suitable for medium voltages. Two opposed disks or plates, one fixed and one free to move axially, opposed disks or plates, one fixed and one free to move axially, are forming common very high voltage instrument. Both direct and alternating r.m.s. values have been indicated, but in general the type has been less robust than others and so it has limited use. Its great advantage for d.c. use is that no current is taken from the circuit. On alternating current, the small current taken is a capacitance current having no power component

Electrostatic Lens : See electron lens.

Electrostatic Precipitator. : It is an apparatus which is used for the removal of suspended matter out of a gas stream. The gas to be cleaned has been passed through an electric field which is established between-so-called active or discharge electrodes and receiving electrodes. The active electrodes (usually wire) have been maintained at a high negative direct voltage (10—100 kV) and take a current of about 10—100 mA. The high electric field strength in the vicinity of the wire causes electric breakdown of the gas. Negative ions and electrons has been thus created near the wire, travelling towards the receiving electrodes and charging the dust particles negatively. The charged particles have been electrostatically attracted by the receiving electrodes and deposited on them. Rapping mechanisms of various types have been fitted to dislodge the collected dust which then drops by gravity into a hopper.

Electrostatics : Refers to the study of the behaviour of electric charges and potentials. The basis phenomenon of electrostatics has been the force exerted between two charged bodies.

Electrostriction : Electrical phenomenon which is analogous to magnetostriction. Polycrystalline barium titanate and lead zirconate have been electrostrictive materials, and vibrate when subjected to an alternating electric field.

Electrotherapy : Refers to the treatment of diseases and other physical disorders by electric currents or by electrically produced radiation. It is also termed as physiotherapy.

Electrothermics : Refers to the application of electric energy to the production of heat in chemical or metallurgical processes.

Embedded Temperature Detector : It is a thermocouple or reisitance temperature type of temperature-sensitive device which is built into an electric machine for monitoring hot-spot temperatures.

E.M.F. : See electromotive force.

Enamel : A thick oil point having resin and having good insulating qualities. It has been extensively used for covering fine wires.

Encapsulation : Refers to the encasing of electric components in soild blocks of resin. Polyester or epoxy resins have been used, the latter giving better results. The procedure is sometimes termed as potting.

Enclosed-ventilated Equipment : Equipment in which the live parts get protected by covers having opening for ventilation.

Enclosure : Structure having an electric machine.

End Bracket : Refers to an open structure at the end of the frame of a machine to support a bearing.

End Cell : Alternative name for regulator cell.

End Plate : Refers to a thick plate between two of which the laminations of a laminated structure have been clamped.

End Shield : A cover which is wholly or partially enclosing the end of a machine and posibbly equipped to support a bearing. It is different from an end guard, which does not carry a bearing.

End Spring : Refers to a spring of hard lead which is kept in a lead–acid cell between the outer negative plates and the end of the container to prevent the plates from spreading.

End Winding : Refers to the part of a winding which is lying beyond the slots, serving simply to connect the two active coil sides and contributing nothing to the output of a machine. It is also termed as an overhang.

Endurance Test : It is a specified test which is used to ascertain th capability of a device to endure without any detriment particularly severe operating conditions like vibrations, shocks, shot-circuits, over voltages or over-currents.

Energy :

(1) The capacity of a system to do work. The SI unit is the joule (symbol J). Supplementary units have been the kilowatt-hour and the electron-volt.

Energy Component : Obsolete name for active component.

Energy Level : A term which described the state of axcitation of an electron in an atom or molecule.

Envelope : It is the gas-tight container of an electronic tube or valve, which is made of glass, ceramic, metal or a mixed construction.

Epoxy Resin : A synthetic thermosetting resin. Mixed with suitable fillers (*e.g.* glass fibre or asbestos), or applied to sheet material, it will give castings or laminates of varying degrees of electric and machanical strength, sometimes in very large pieces not readily made by ordinary moulding methods. It is essentially useful for high-voltage power switchgear insulation.

Epstein Square : Refers to an assembly of strips of core-plate material which is arranged in the form of a hollow square interleaved at each corner. Each limb has been provided with primary (exciting) and secondary windings for the measurement of core loss.

Equaliser : Refers to the connection between two points of a winding, so that the potential at the two points has been the same. The electromotive force produced in one path of a lap-conected d.c. armature winding may differ from that in another because of magnetic asymmetry. Even small differences may produce large internal circulating currents, with consequent sparking at the brushes and excessive heating of the winding. These effects could be mitigated by using a sufficient number of equaliser rings of low resistance, which connect together all points on the winding that, ideally, should have no difference of potential between them. These may also be termed as equipotential connections.

Equipotential Connection : Alternative name for an equal- iser.

Equipotential Surface : Refers to a surface within which there exists no potential difference between any points.

Equivalent Circuit : It is an electric circuit which is set upon to represent an energy system, electric or otherwise. It gives results in terms of measurable (or calculable) voltages, currents, charges, linkages, etc., that can be employed to represent analogous quantities in another physical system or network.

Equivalent Conductance : May be defined as electrolytic conductance of one gram-equivalent of a substance when a potential difference 1 V exists between electrodes 1 cm apart. It has been equal to the conductivity multiplied by the number of millilitres per gram-equivalent. Maximum equavalent conductance is applicable to a solution when this is at infinite dilution in its own solvent.

Equivalent Generator : Device used in circuit analysis. An electric energy source (such as a microphone, a battery, a synchronous generator or a valve amplifier) could be represented by an equivalent generator comprising a source of e.m.f. E_0 and an internal impedance Z_0 in series with it. On occasion (*e.g.* with pentode valves) it has been more read to regard the source as producing a constant current I_0 with an internal admittance Y_0 connected accross it.

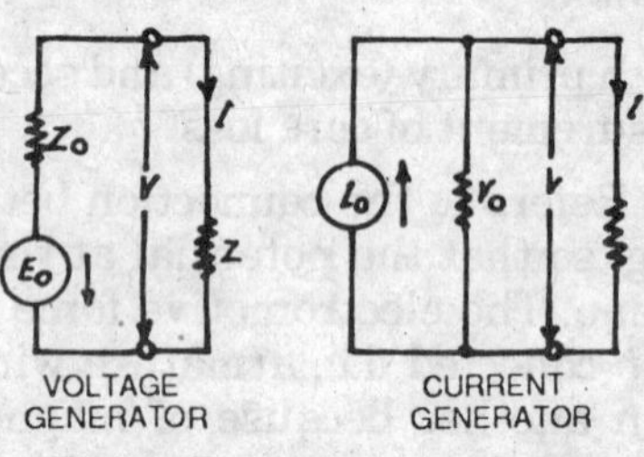

Fig. 4. Equivalent generator.

Equivalent Resistance : Refers to the real part of the impedance complexor of a circuit element. For an element in a cirucit with sinusoidal current the value of its equivalent resistance could be termed act by dividing the dissipation caused by the current by the square of its r.m.s. value. The value of equivalent resistance of the emelemt is greater than that of its d.c. resistance because of a.c. phenomena such as eddy current and skin effect.

Equivalent Sine Wave : It is a sine wave which is having the same r.m.s. value and fundamental frequency as the given wave.

Ettingshausen Effect : Refers to the appearance of a temperature gradient in a single material because of the interaction of an electric current and a magnetic field.

Eureka : It is the trade name for cupro-nickel alloy which is used in the manufacture of resistance wire. It is having a small temperature coefficent and will withstand high temperature.

Excitation : Refers to the magnetomotive force which is producing the magnetic flux in the field system of an electric machine or in an electromagnet or in a transductor. The term is applicable to the force itself or to the effect produced.

Exciter : It is an equipment which is used to provide the field current of electric generators. Most exciters have been rotating machines, although on occasion rectifiers have been used for converting a suitable variabale-voltage a.c. supply to direct current. As it is unusual to include a variable resistance in the main-field circuit, any change in excitation current is obtained by adjusting the applied voltage. Consequently it must be possible to vary the exciter voltage over a wide range.

Excitron : It is a single-anode mercury-arc rectifier. It is made in both glass and steel, but is usually pumpless.

Exploring Coil : A coil which is used for measuring magnetic field flux.

Explosion Pot : It is a small vessel of insulating material which is surrounding the fixed contact of an oil circuit-breaker, its outlet being sealed by the moving contact. An arc in this pot produces high gas pressures, and when the moving contact clears the outlet the arc gets extinguished by a blast of oil and gas.

Explosion Vent : Refers to a frangible diaphragm which is fitted to a transformer, generally in lid, to save the tank from bursting in the event of an internal explosion that might result from an electric fault. It has been often shrouded with a pipe to throw escaping oil clear when it operates, and varies in diameter with the size of the transformer, from about 15 to 30 cm.

Expulsion Fuse : It is a fuse in which the extncition of the arc gets facilitated by an increase in length of the break in the link. This is attained by expulsion of part of the fusible material.

Expulsion Gap : It is a surge diverter in the form of a type of expulsion fuse which may get connected in series with a gap across insulator strings of an overhead transmission line. On a voltage surge, expulsion of the hot gases and ions effectively may interrupt any follow current and does not allow disruption of the supply. The device is sometimes termed as a protector tube.

Extra-low Voltage : A term used in European regulations to imply a voltage not exceeding 50 V.

Extrinsic Conduction : Refers to a process of impurity conduction which is occurring in semiconductors, where it predominates over intrinsic conduction. Extrinsic conduction gets resulted from the presence in the base material of a small number of atoms of other elements. The foreign element could be a donor, in that it gives one possible free electron per atom to increase conduction and to yield n-type properties. Alternatively, an acceptor element may give rise to one possitive hole per atom, endowing the base material with p-type properties.

Extrusion : The term used for the production of rods, tubes, etc., by forcing hot material through a suitbale die by means of a ram. Extrusion could be used for sheathing

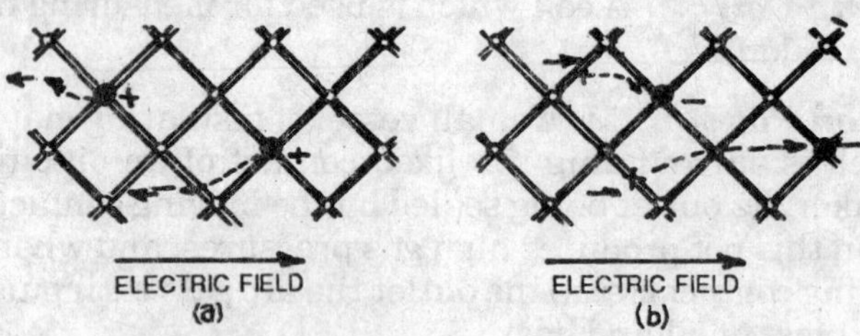

Fig. 5. extrinsic (impurity) conduction. Donor impurities (a) cause excess-electron (n-type) conduction. acceptor impurities (b) cause hole (p-type) con-

electric cables, when the temperature used gets limited by the presence of the insulating papers.

Faceplate Breaker Starter, Controller : It is a faceplate starter, or controller, which includes an interlocked contractor for removing the task of breaking or interrupting the circuit from the starter or controller contacts.

Faceplate Starter, Controller : It is a starter or controller which is having the switch contact part's arranged upon a plane surface.

Fahy Simplex Permeameter : It is a yoke permeameter which does not have compensation. A bar specimen gets clamped between the end of a single rectangular yoke that carries the magnetising winding. The search-coils for measuring B and H ballistically have been extending along the whole length of the specimen, the H-coil being mounted between two massive soft-iron posts clamped to the ends of the specimen.

Farad : It is the SI unit for capacitance (Symbol : F)

Faraday : A unit used in electrochemistry which may be defined as the quantity of charge required to deposit one gram-equivalent of any ion. It is equal to 96 487 coulombs.

Faraday Cage : Refers to a closed box having a conductive surface which serves, so far as electric effects from charges at rest get concerned, to divide the universe into two independent parts–the part within the box and the part outside. It could be constructed adequately of wire netting so long as the instruments or persons to get screened and the charges setting up the field have been not closer to the netting than about ten times the mesh-opening. Whether of wire mesh or of a continuous unbroken conductive surface, the Faraday cage would not be able to screen static or slowly changing magnetic fields. It will, however, reduce the intensity of electromagnetic waves if their wavelength has been appreciably longer than the width of the mesh-opening.

Faraday—Neumann Law : It is a law of electromagnetic induction. Faraday proved than an electric current gets induced in closed circuit when a magnet is moved near it. The induced current can be produced also by moving the circuit near the magnet, by changing the circuit area, and by changing the magnetic field intensity. A more comprehensive statement by Neumann has been as follows : An electromotive force would be set up in a circuit when the magnetic flux linking the circuit gets changed in any manner. the magnitude of the e.m.f. has been proportional to the time rate of change of flux-linkages with the circuit, and is having such a direction that any current it produces tends to oppose the change of flux-linkage. The law may be expressed.

$$e = -d\psi/dt$$

where e denotes the total e.m.f. and ψ the flux, or the effective flux tern product. It has been often possible to use the relation $\psi = N\phi$, ωηερε ϕ denotes the average flux per turn, and N the number of turns of the circuit. This is the more familiar engineering approach.

Faraday's Laws of Electromagnetic Induction : A law which was derived basically by Faraday, and in a more comprehensive form by Neumann.

Faraday's Laws of Electrolysis : During the flow of an electric current through an electrolyte, the mass of ionic material set free at either electrode, deposited thereon, or dissolved therefrom, has been proportional to the quantity of charge that passes and the ionic weight of the ion and inversely proportional to the ionic valence of the material being electrolysed. It has been found to be independent of other factors such as the voltage, electrode size current density and temperature.

Faradism : Electromedical treatment which is used a pulsed current, derived from an induction coil, for stimulating muscles and nerves.

Fault Current : Refers to the current which is flowing from one conductor to another or to earth, because of a fault in the insulation.

Faure Plate : It is a plate of a lead-acid cell which is made mechanically by the application of lead-oxide paste on to a grid. It is also termed as a pasted plate.

F-class Insulation : Refers to one of the seven classes of insulating materials for electric machinery and apparatus defined on the basis of thermal stability in service. It is assigned a temperature of 155°C and is involving materials such as mica, glass fibre or asbestos suitably impregnated or coated.

Fachner Law : If brightness appears to get increased in equal steps, it is actually being increased logarithmically. The minimum fractional difference in brightness that can just be detected is termed as Fechner's fraction.

Feedback : Refers to the injection of a fraction of the output signal or response of a system into the input which has been externally excited by an input signal or stimulus. The feedback has been positive (regenerative) or negative (degenerative), depending on whether a cophasal or antiphasal component has been present in the fed-back signal.

Feedback Oscillator : Refers to a tuned feedback amplifier in which the feedback voltage an amplitude and phase such that sufficient positive feedback is produced for oscillations to be sustained.

Feeder : A line which is supplying a point of a distributing netwrork, and not having any intermediate connections.

Feeder Box : Alternative name for a junction box.

Feeder Pillar : A pillar having switches links or fuses for connecting a feeder with distributing mains.

Feet Switch : Alternative name for a tropical switch.

Ferranti Effect : Refers to a transmission-line phenomenon. In a line having length substantially less than one half-wavelength but long enough to have appreciable shunt capacitance,the voltage across the receiving end on no-load (or open circuit) might be greater than that applied at the sending end. The effect has been the result of electric and magnetic energy interconversion imposed by the open circuit conditions at the receiving end.

Ferrimagnetism : It is a phenomenon which is exhibited by materials in which two different kinds of ion, one having a greater magnetic moment than the other, are in opposition. It is usually displayed by a ferrite.

Ferrite : A crystalline oxide is having the general formula Fe_2O_3. MO Where M is a divalent metal. Some (*e.g.* M = zinc) exhibit antiferromagnetism; other *e.g.* M=Copper, nickel, iron) exhibit ferrimagnetism. All have been less dense than iron and behave as 'soft iron' materials. They find use in transformer cores, As they have very high resistivity, their eddy-current losses a high frequencies have been small compared with those of metals.

Ferrodynamic Instrument : It is an electrodynamic instrument in which the action gets augmented by the presence of ferromagnetic meterial.

Ferroelectric : Applicable to some dielectries which are exhibiting the phenomenon of electric hysteresis, analogously as a ferromegnetic material exhibits magnetic hysteresis.

Ferromagnetic : Applicable to materials exhibiting ferromagnetism.

Ferromagnetism : A phenomenon which is exhibited by materials having a relative permeability which is considerably greater than unity, and which varies with the flux density. Examples of such materials have been iron, steel, cobalt and nickel.

Ferro-resonance : Refers to a condition of resonance in a series inductance capacitance circuit, the indcutor having a ferromagnetic core. The combined current/ voltage characterisitc will exhibit a point of instability at which a change in voltage makes a discontinuous jump in the characterisitc. Ferro-resonance may take place with an unloaded transformer connected to a cable system, or it may get exploited for relay operation.

F.E.T. : Field-Effector Transistor.

Feussner Potentiometer : It is a convenient form of practical potentialmeter which is involving five or more dials. The overal resistance remains constant at all settings, because three double-contact dial switches take out resistance from one circuit as they add an equal resistance to the other. Two single-contact dial switches have been connected to the unknown test circuit, through a galvanometer in series, as is seen in Fig. 1. The reading could be done directly from the dials, which are in decades.

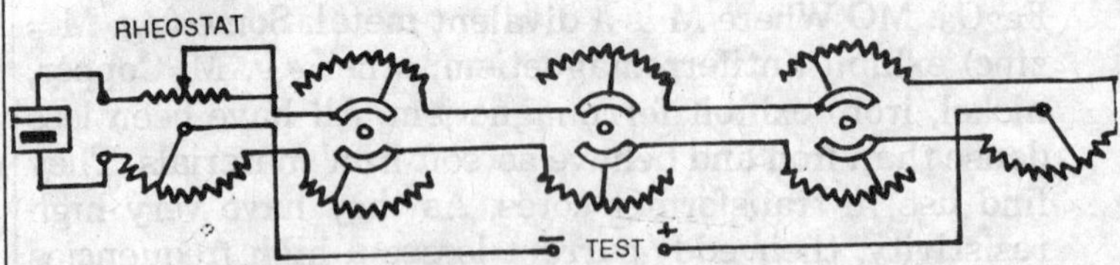

Fig. 1. Principle of Feussner potentiometer

Field : A concept regarded as to provide a basis of explanation for some physical phenomena which might otherwise be which was designed representing action at a distance. A scalar field is representing the distribution in space of a scalar quantity, such as temperature (which has a magnitude but no directional properties). A vector field is describing the distribution in space of the direction and magnitude of a vector quantity, such as velocity, current flow, electric force, magnetic flux density, or head flow. It is possible to map such fields in terms of flowlines which give tne direction at various points of the flux function concerned, and which by their concentration give some idea of the density of flow. In other words, a tield in a given region of space involves the aggregate of physical quantities that are functions of position within the region. Many fields have been conforming with the Laplace equation

$$\frac{\partial^2 \phi}{\partial x^2}+\frac{\partial^2 \phi}{\partial y^2}+\frac{\partial^2 \phi}{\partial z^2}=0$$

where ϕ denotes the flux function, and xyz have been the coordinates of space.

Field Coil : Current-carrying coil which is used to energise the field magnet of a machine.

Field Control : It is a method of motor speed control in which the excitation gets varied by shunting the series winding, cutting out some series turns, or inserting resistance into the shunt field circuit.

Field-discharge Switch : It is a switch which is used for controlling the field circuit of a motor or generator. It gets connected in series with the field windings and is having additional contacts so that a discharge resistor gets connected across the windings when the supply to the field circuit is interrupted. It is also termed as a field-breaking switch.

Field effect Transistor (F.E.T) : It is a voltage-controlled semi-conductor device having a source region and a drain region separated by a gate region (three electrode f.e.t.) or by two independent gate regions (four electrode f.e.t.). The gate voltage is able to control the conductance of the thin layer, called the channel, between source and drain regions and thus the rate of flow of charge carriers. The device is having a high input impedance and a low noise figure.

Field Emission : Alternative name for cold-cathode emission.

Field-failure Relay : It is a motor protective device which opens a shunt motor starter should the motor field collapse. Its application has been generally restricted to drives where loss of the field would results in a dangerous condition such as over-speeding or the loss of dynamic braking. The operating coil of the relay gets connected in series with the field, while its normally open trip contacts have been the undervoltage release circuit.

Field Magnet : Refers to the electromagnet or permanent magnet that gives rise to the magnetic field in an electric machine.

Field Regulator, Rheostat : A variable resistance which is used for varying the current in the field winding of an electric machine.

Field Spider : It is a rotating spoked structure which carries, at its periphery, the pole pieces of an electric machine.

Field Spool : The structure on which a field coil has been carried. The term is sometimes used to refer to the field coil itself.

Field Suppression : Refers to the reduction of the field of a generator in the shortest possible time in the event of an internal fault. The field could be reduced to zero by disconnecting the main field from its exciter and discharging it through an appropriate resistor, by open-circuiting the field winding or by exciter reversal.

Field's Test : It is a test for a series motor which is similar to a back-to-back test, but having the generator output dissipated in a resistance instead of returned to the motor. The generator field has been in series with the motor for ensuring that the generator always excites itself.

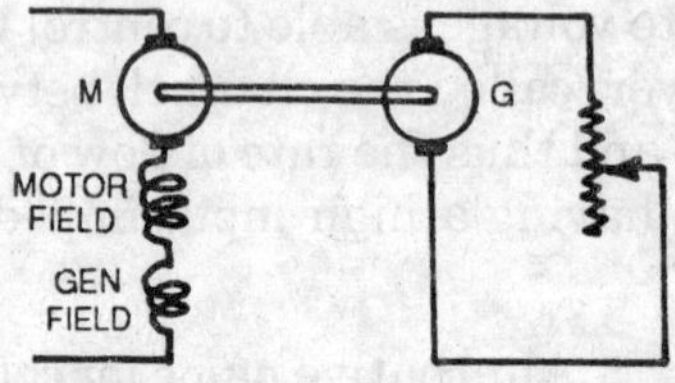

Fig. 2 Field's test

Filament : In a lampt it is the thread-like conductor which is made incandescent by the passage of an electric current. It is gnerally made of a metal such as tungesten.

Filament Lamp : It is a lamp having a fine filament in an evacuated or gas-filled glass envelope.

Filter (Network) : In a.c. network which has been so designed that it passes energy at frequencies within one or more bands and prevents the passage at all other frequencies.

Fisher Loop Test : It is a modified loop test that uses two sound conductors which may be of unknown lengths and different cross-section from the faulty conductor. In Fig. 3 if a_1 and b_1, a_2 and b_2 have been the values of the resistances for balance in the appropriate positions of the switch, and x has been the distance of the fault along the cable.

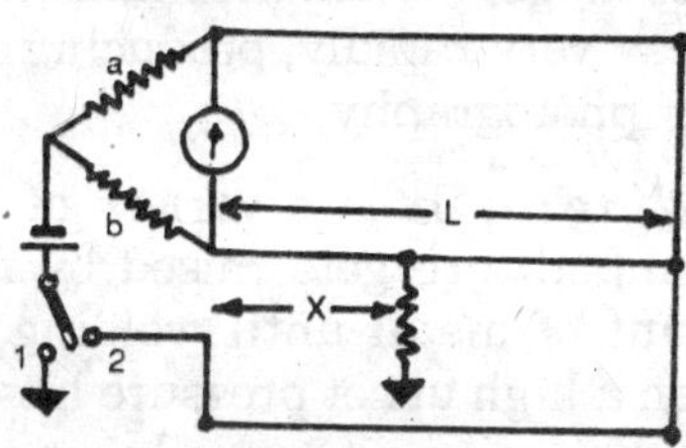

Fig. 3. Fisher loop test for a fault in a conductor (Length L).

$$x = L \frac{b_1 (a_2 + b^2)}{b_2 (a_1 + b_1)}$$

Fixed Loss : Refers to the sum of the core loss, windage, bearing friction and brush friction of electric machine, taking place with no load on the machine. It does not include the exciting-circuit or exciting-current loss.

Fixed-trip Circuit-breaker, Switch : It is a curcuit-breaker or switch, the automatic release mechanism of which cannot function independently of the closing mechanism.

Flame-cored Carbon : It is a type of cored carbon. The channel fillings have been a mixture which colours the arc.

Flameproof : A term which is applicable to enclosures to signify ability to withstand, without injury, any explosion of a specified flammable gas that may take place within the enclosure under normal conditions, and ability to disallow ignition of the gas in the surrounding atmospher.

Flammable : Means that the relevant substance or material can easily get ignited. In contrast non-flammable indicates that the substance or material has been difficult to ignite.

Flash Barrier : A screen which is made of insulating and non-flammable (see flammable) material used to inhibit arc formation, or at least to reduce the effects of an arc.

Flash Bulb : It is a transparent envelope usually having a metal foil in contact with a low-voltage filament in an atmosphere of oxygen. When the filament gets heated the foil burns away very rapidly, producing a high-intensity flash of use in photography.

Flash Butt Welding : It is a variety of butt welding in which the temperature gets raised by flashing away a certain amount of metal until welding temperature is reached, when a high upset pressure has been applied.

Flashing Fault : It is a fault which is not evident at low voltages, but which makes repeated flashover when a higher voltage is applied.

Flashover : Refers to the accidental occurrence of an arc between two conducting parts of a machine or apparatus.

Flashover Test : A test which is applied to equipment designed to operate at high voltages. The voltage gets raised until flashover of the test object occurs. The measured flashover voltage must be not less than a specified minimum vlaue. Both dry and wet flashover voltage-test could be specified, and they usually follow on consecutively from the appropriate withstand test.

Flat-compounded : A compound-wound generator in which a series winding on the field system, connected to act cumulatively to the shunt winding, has been used to neutralise the voltage drop which would otherwise occur as the load builds up. It is also known as level-compounded.

Flat Pressure Cable : An oil-filled cable which is having three cores laid side-by-side and sheathed with lead. The lead gets reinforced with metal tapes so that the flat side of the sheath can act as a loaded beam. The cable has been useful for submarine cables, and for land cables up to 66 kV and occasionally 132kV.

Flatrate Tariff : A tariff which is involving a single charge proportional to the number of units of energy supplied.

Fleming's Rules : A method which is used for indicating the relative directions of current, mechanical force and magnetic field in the case of a current lying in a magnetic field. The thumb and forefinger and second finger get arranged roughly at right-angles in three directions corresponding to co-ordinate axes. Then the Forefinger gives the direction of the Field, the second finger that of e.m.f. or current, and the thumb provides the direction of motion under action of the force. With the left hand, the directions have been then indicated for motor action; with the right, for generator action.

Flit Plug : A detachable cable sealing box for attaching to apparatus and coupling cables.

Floating Battery : A battery which is connected in parallel with a direct-voltage supply whose normal voltage has been a little higher than the open-circuit voltage of the battery. A small charging current then flows to maintain the battery in a fully charged condition.

Floodlight : High-intensity illumination of the whole of a scene or object by positioning a lamp at the focal point of a reflector or lens system. A bunch filament provides the compact light source necessary.

Floor Warming : Refers to a form of space heating in which the heat has been applied to concrete floor mass. The two principal types have been (a) withdrawable, in which a heating cable rendered waterproof by suitable sheathing has been drawn into a tube itself embedded in the concrete, or a preformed hole in the floor; heat has been conveyed to the concrete mass partly by conduction, and partly by convection through the air heated in the tube, and (b) directly embedded, where the heating cable has been itself completely embedded in the concrete, heat dissipation being by conduction; because of the direct contact with the mass of the floor an even heat distribution is maintained, and the temperature of the heating cable is less than with the withdrawable system.

Fluorescent Lamp : It is a lamp in which gas conduction through low pressure mercury vapour has been used to generate ultraviolet radiations, which in turn are changed into light rays by the fluorescent effect of a coating on the inside of the glass tube. By suitable choice of the chemical coating, the quality of the light output could be varied.

Flux-Cutting Law : If a conductor, of length l, gets moved at right-angles to a magnetic field of uniform density B at velocity v, the electromotive force induced has been as follows

$$e=Blv$$

Flux-Diversion Relay : Refers to a polarised relay in which a permanent flux exists. The application of operating current gives rise to a flux which counteracts the permanent flux, diverting it to a magnetic circuit which includes the armature and an air-gap. The armature has been thereore attracted and has been held by the permanent magnet flux even if the current is subsequently disconnected.

Flux-linkage : Refers to the product of the magnitude of mangetic flux and the number of turns of a coil or circuit through which they pass. The unit of Flux-linkage has been l Wb-turn (weber-turn).

Flexmeter : It is an instrument that measures magnetic flux by using a search coil that may get moved into or out of a magnetic field, the induced quantity of charge in flux-turns being fed to a moving-coil instrument. The fluxmeter has been a galvanometer in which the controlling torque has been made as low as possible by employing a suspension or ligament above and below the coil of soft metal, not phosphor-bronze as has been normal for gal-

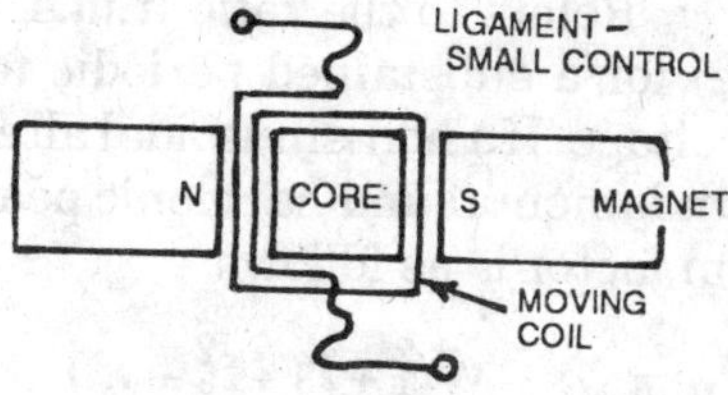

Fig. 4. A simple fluxmeter

vanometers. In addition, the damping has been made high by the use of a relatively thick aluminium former on which the coil has been wound, swinging in the permanent-magnet field (see Fig. 4). A long period has been thereby attained.

Follow Current : Refers to the current that succeeds a high-voltage surge through a discharge path after the discharge is initiated.

Foot-candle : Unit for illuminance. It has been now superseded by the SI unit lux. 1 foot-candle=10.764 lux.

Foot-lambert : Unit for luminance. It has been now superseded by the SI unit candela per square metre. 1 foot-lambert=3.426 candelas per square metre.

Forced-circulation Boiler : It is a boiled which is having pumped water flow to overcome problems of stream and water separation that may be encountered in high-pressure generating plant.

Forced Cooling : Refers to the removal of heat from a device or equipment or plant by employing mechanical means to circulate a coolant.

Forced draught Ventilation : Refers to a form of machine ventilation in which the ventilating air has been supplied under pressure by some means external to the machine itself.

Forced Oscillation : Refers to a condition which is established when a system inherently oscillatory is energised by means of a periodic driving function.

Form Factor : Refers to the ratio (r.m.s. value/half-cycle mean value) for a substained periodic function such as current or voltage. If a non-sinusoidal alternating current is having fundamental and harmonic peaks of i_1, l_3, i_5,.... then its form factor is as follows.

$$K_f = \frac{\pi}{2\sqrt{2}} \quad \frac{\sqrt{(i_1^2 + i_3^2 + i_5^2 + \ldots)}}{i_1 + \frac{1}{3} i_3 + \frac{1}{4} i_5 + \ldots}$$

For a pure sine wave, $K_f = \pi/2\sqrt{2} = 1.11$.

Formative Lag : In a charged spark gap, this terms refers to the time between the occurence of an electron in a position suitable to initiate breakdown and the actual passage of the spark.

Formed Plate : It is a plate of a lead-acid cell which is prepared by electrolytic conversion. it is also known as a Plante plate.

Former : A tool which is used for forming a coil or winding into the required shape.

Foster Reactance Theorem : It is a theorem which may be applied in circuit synthesis to provide a required reactance/frequency relation. In a general network of pure reactances between two terminals, the impedance at the driving point takes the following forms.

$$Z = j\,\omega H \frac{(\omega^2 - \omega_2^2)\,(\omega^2 - \omega_4^2)\ldots}{(\omega^2 - \omega_1^2)\,(\omega^2 - \omega_3^2)\ldots}$$

$$Z = \frac{H}{j\omega} \frac{(\omega^2 - \omega_1^2)\,(\omega^2 - \omega_3^2)\ldots}{(\omega^2 - \omega_2^2)\,(\omega^2 - \omega_4^2)\ldots}$$

where H has been a positive constant, and $0 \leq \omega^1 \leq \omega_2 \ldots < \infty$. The expressions give the variation of impedance with the frequency ω at which the network is supplied.

Four-terminal Network : This term has been now superseded by two-port network, with the alternative term two terminal network.

Fourier Series : A series used in harmonic analysis. If a finite, univalued function g(x) recurs, may be regarded to recur, over successive intervals of 2π in the vaiable x, so that $g(x)=g(x+2m\pi)$ for all integral values of m, then it has been possible to represent it by a synthesis of sinusoidal waves of fundamental period 2π and of various integral harmonic orders.

Thus, a function $y=g(x)$ could be written as either

$$y=a_o+a_1 \cos x+a_2 \cos 2x+...+a_n \cos nx+...$$

$$+b_1 \sin x+b_2 \sin 2x+...+b_n \sin nx+...$$

or

$$y=a_0+c_1 \sin (x+_1)+c_2 \sin(2x+\alpha_2)+...+\alpha_n+...$$

Each of the expressions above has been a Fourier series, where

$c_1=\sqrt{(a_1{}^2+b_1{}^2)}$, $c_n=\sqrt{(a_n{}^2+b_n{}^2)}$ have been the amplitudes of the fundamental and nth harmonic, and the phase angles are $\alpha_1=\arctan(a_1/b_1)$, $\alpha_n=\arctan(a_n/b_n)$, and a_o is a constant.

Fractional-pitch Winding : A type of winding that has been accommodated in a number of slots not divisible by the product of the number of phasis and poles. It is having fractional value for number of the slots per pole, and is also known as a fractional-slot winding. For the purpose of reducing harmonic e.m.f.s., an armature winding may be having a span greater or (more usually) less than a pole-pitch. In a three-phase machine the coil-span gets arranged generally to reduce the fifth and seventh harmodic e.m.f.s. the third, ninth, etc. being suppressed by the three-phase connection.

Frame Fault Protection : It is an earth-fault busbar protection scheme which is applied to metalclad switchgear, the arrangement shown in Fig. 5 being for a two section switchboard. In order to afford protection, the switchgear framwork has been lightly insulated from the ground. It has been then earthed through a current-transformer primary conductor. The arrangement shown discriminates between faults on the two busbar sections, insulated barrier has been kept between the bus-section switch framework and the righthand section framework.

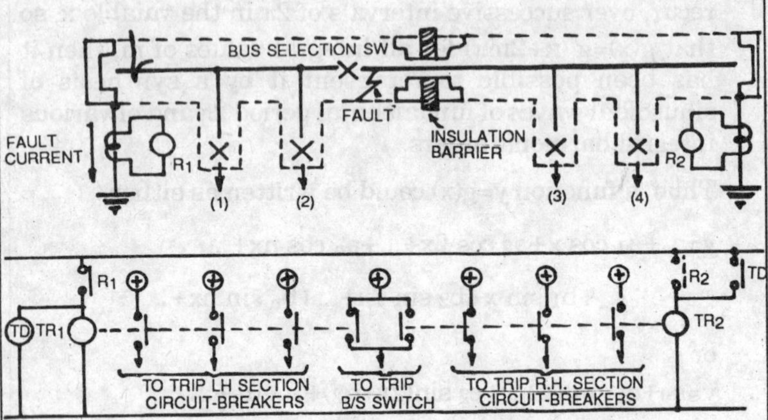

Fig. 5 Connections for two-zone frame fault protection

The two framework have been separately earthed, faults in the left-hand section initiating operation of relay R1, and in the right-hand section R2. For high-speed fault clearance of all faults, insulation of the switchgear framework has been provide on both sides of the bus-section framework has been separately earthed. The bus-section framework has been then tripped for current in any of the three earth connections, and current in the bus-section earth which is arranged to initiate tripping of both of the main sections.

Francis Turbine : It is a reaction-type water turbine which is used for medium heads of water up to approximately 300m. The water is allowed to flow through a spiral casing inwards through the wheel and acts on a series of blades running between the hub and an outer wheel ring. Use has been thus made of both the kinetic and the potential energy in the water. The complete installation is including a casing with a ring of movable guide vanes, round the circumference of the runner. Mechanical linkages, operated by servomechanisms, have been operating these vanes to regulate the flow of water to the turbine.

Frequency : May be defind as the number of complete cycles of a periodic function (*e.g.*,of current or voltage) in one unit of time. Its SI unit frequency has been the hertz (Symbol : H_2).

Frequency Changer : Term used in the USA for frequency converter.

Frequency Converter : See conver.

Frequency Distortion : Refers to change in waveform, due to the input components being altered in their relative magnitude by a frequency-dependent reponse in the system. This is also termed as attenuation.

Frequency Modulation : Refers to the imposition of a signal on to an alternating carrier wave in such a way that the amplitude of the carrier remains constant while its frequency has been varied proportionally to the amplitude of the signal.

Frequency Transformer : It is a transformer which is used for converting alternating current at one frequency to alternating current at another. This term should always be used where a static transformer has been used for this purpose without a rotating machine, to distinguish its from a frequency changer.

Frosted Lamp : It is an electric lamp which is having its bulb partially or wholly etched or sandblasted to diffuse the light. An internally frosted bulb has been the principal feature of a pearl lamp.

Fulchronograph : Refers to an instrument which is used for measuring the variation with time of the current in a flash of lightning. The latter has been permitted to energise an electro-magnet, between the poles of which move magnetic surge-current indicators.

Fundamental Component : Refers to the component of fundamental frequency in the harmonic analysis of a non-sinusoidal quantity.

Fuel Cell : A device which is used for converting the chemical energy of fuels into electric energy without going through the intermediate stages of heat and mechanical energy involved in thermal generating plant. Two typical fuel cells have been depicted in Fig. 6. The first of these is an H_2-O_2 cell (hydrox cell) of the type developed by Bacon, operating at 200°C and 4MN/m^2 where as the second uses air and almost any gaseous or vaporised fuel

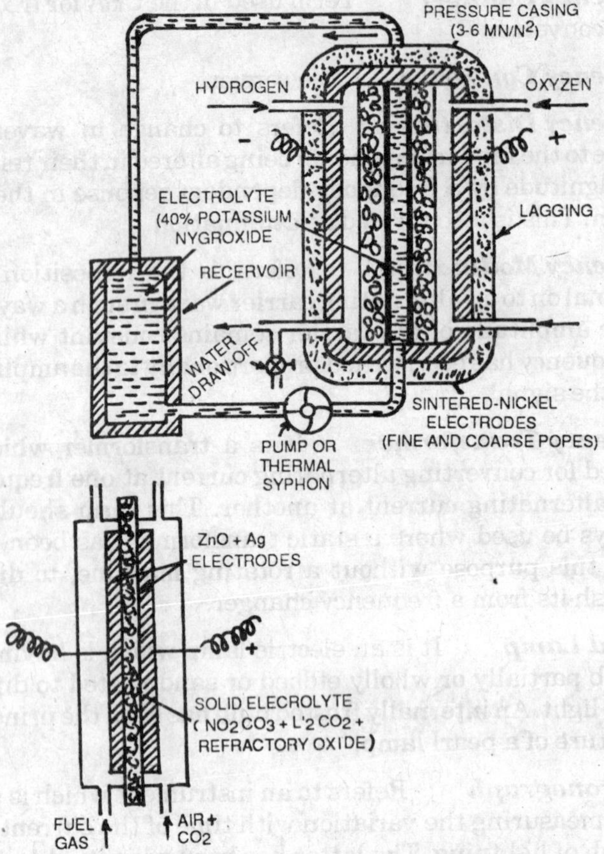

Fig. 6 Fuel cells : above is a hydrogren-oxygen cell; left is a high-temperature gas cell

and operates at about 550-700°C. There has been considerable interest in the devleopment of carbox cells, which would burn liquidhydrocarbon fuels. Fuel cells are irreversible.

Fundamental Factor : May be defined as the ratio of the r.m.s. value of the fundamental component of a non-sinusoidal quantity to the r.m.s. value of the quantity itself.

Fuse : It is a device that is able to protect a circuit against damage from excessive current flowing in it by the melting of a fuse element. When the current melts the fuse element, the circuit has been opened. The fuse includes all the parts that form the complete device.

Fuse Element : Refers to that part of a fuse that has been designed to melt when the current exceeds some specified value for a specified time.

Fuse Link : Refers to that part of a fuse that contains the fuse element and has been designed to be replaced after the fuse has operated.

Fuse Switch : Strictly, a switch having one or more fuses on its moving part. However, with high voltage equipment, the term is used loosely to cover any combination of a fuse and switch.

Fusing Factor : The ratio which is greater than unity of the minimum fusing curremt to the rated current.

Gall Potentiometer : It is a.c. potentiometer that gives two variable components of electromotive force which are in quadrature with one another. These components have been summed and balanced against the unknown e.m.f. In Fig. 1, the supply voltage E gets applied to the primaries of the isolating transformers T_1 and T_2. The secondary of T_1 has been supplying a current, approximately in phase with the supply voltage to R_1, which has been consisting of a tapped resistor and slide wire in series. The current in this circuit could be adjusted by the rheostat R_3. The current also passes through the primary

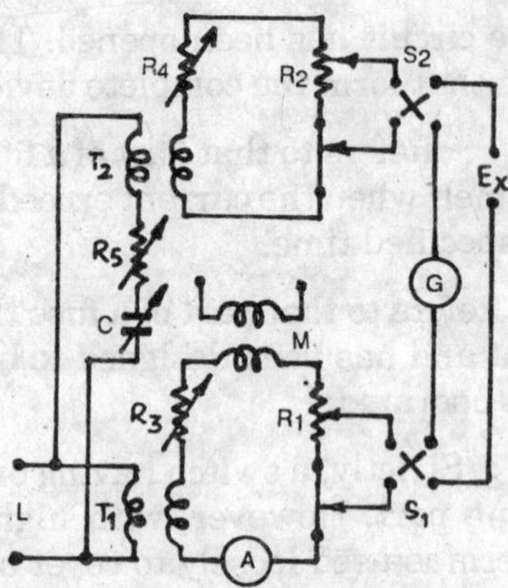

Fig. 1. Gall Co-ordinate potentiometer. The current in R_1 is in quadrature with that in R_2.

winding of the fixed mutual inductor M. Any desired value of in-phase voltage within the range available may be obtained from the tappings on R_1 shown. The primary of transformer T_2 has been provided with a series rheostate R_5 and a variable capacitor C which may be so adjusted that the current in R_2 has been in quadrature with the current in R_1. These components in series are then balanced against the unknown voltage E through the vibration galvanometer G.

Galloping : The term used for the overhead conductors when they oscillate because of the aerodynamic effect of ice accretions or of their profile. They may also be described as dancing.

Galvanism : Refers to the treatment of diseases by using direct current. It is also known as medical electrolysis.

Galvano-magnetic Effect : Refer's to an effect depending primarily on the fact that an electrically charged particle, moving at right-angles to a magnetic field, experiences a force in direction which is perpendicular to both the field and the motion. The magnitude of this force has been proportional to the velocity of the particle. Examples are the Ettingshausen effect and the Hall effect.

Galvanometer : It is an instrument to which is to used detect or indicate the presence or absence of a small

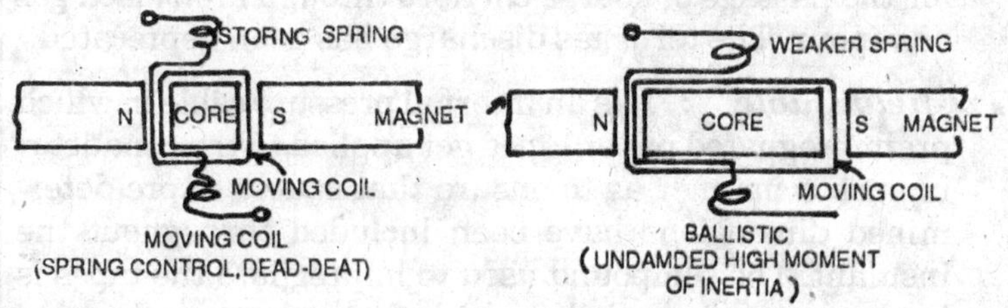

Fig. 2. Moving-coil galvanometers.

current. Most galvanometers have been of the moving-coil type for direct current, high sensitivity being attained by using one or two light phosphor-bronze control suspensions. The two-suspension or bifilar type includes, as well as control by torsion, a form of gravity control, as the coil gets deflected either way from zero. With this suspension the movement is generally shown by means of a mirror on the moving system which reflects a beam of light projected on to it from an external lamp. The ballistic moving-coil galvanometer possesses a small controlling torgue and minimum damping it is able to read impulses in flux-turns or microcoulombs and the reading is taken at maximum deflection. In order to get accuracy the discharge must occur into the coil quickly, before the movement commences to deflect.

Gap **:**

(1) Refers to the space (normally air-occupied) between the stationary and movable parts of an electromagnetic machine.

(2) Also refers to the distance between the electrodes of a high-voltage measuring or limiting device like a sphere or rod gap.

Gas-and Pressure-actuated Protective Relay : Alter na-tive term for Buchholz relay.

Gas-blast Circuit-breaker : Refers to a high-power circuit-breaker in which a blast of gas is allowed to accross the contacts at the instant of separation to extinguish the arc. Recently sulphur hexafluoride is used as an insulant and arcquenshing medium.

Gas Conduction, Gaseous Conduction : The term used for the passage of charge carriers through an ionised gas or vapour. The term gas discharge has been deprecated.

Gas-field Cable : It is an internal pressure cable in which preimpregnated paper tapes get applied to the conductor in such a manner as to ensure that spaces of pre-determined dimensions have been included throughout the insulant. The compound used to impregnate the tapes is having a petroleum-jelly base so that it does not flow and block the spaces at the continuous operating temperature. Gas-filled cables are used mainly at 33kV.

Gas-or Vapour-filled Tube or Valve : Refers to an electronic tube or valve whose operating charcteristics could be determined by its ionised gas or vapour.

Gas-pressure Cable : A cable which is using gas under pressure to disallow ionisation in the insultant. A cable using external pressure is compression cable; one using internal pressure is a gas-filled cable.

Gas Turbine : Refers to a form of thermal generating plant which is able to provide rapid start and shut-down. it finds application in peakload provision and for standby duties. The simplest form of gas turbine in consisting of a

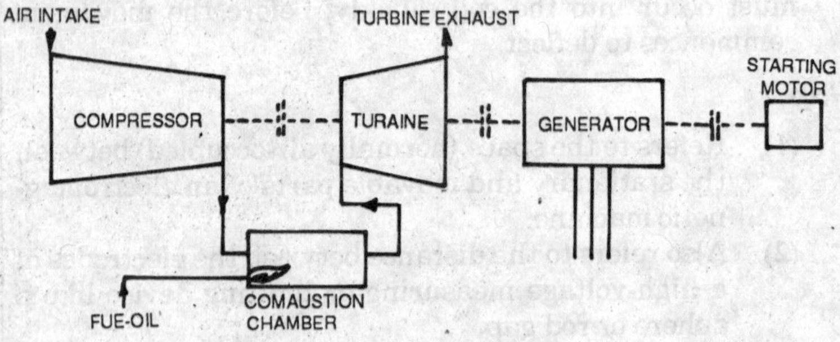

Fig. 3. Simple open-cycle gas-turbine generator

compressor, a combustion chamber and a turbine as shown in Fig 3. Like all turbines, the gas turbine is essentially a constant-speed machine.

Gassing : The term used for the evoluation of gas in an accumulator, which take place near the end of a charge.

Gate-end Switchgear : Refers to an electrically operated contactor switch which is interlocked electrically with a main isolator in a separate compartment. The contactor could be controlled locally or remotely.

Gauss : Refers to the electromagnetic c.g.s. unit of magnetic flux density. The corresponding SI unit has been the tesla. 1 tesla = 1 weber per square metre = 10000 gauss.

Gauss's Law : According to this law, the total number of lines of electric force emerging from a closed surface in an electric field has been Q/ϵ, where Q denotes the total obvious charge enclosed by the surface, and ϵ denotes the permittivity of the insulant in which it lies.

Gaussmeter : An instrument which is used for measuring magnetic field strength at a point.

Gearless Motor : Refers to a traction motor which is having its armature mounted concentrically on the driving axle of an electric locomotive.

Geiger Counter : It is a radiation detector. The Geiger (Geiger-Muller or G-M) counter has been similar in construction to the ionisation chamber, but it has been gas-filled with a high multiplication factor and used at higher voltages so that a pulse of radiation gets indicated or recorded as a pulse of current.

Geissler Tube : Refers to a form of gas-filled tube which works at a moderately low pressure to show the luminous effects of conduction through various gases.

Generating Set : Refers to an equipment which is comprised of one or more generators driven by a prime mover.

Generating Station : Refers to a combination of equipment and its associated housing, for producing electric energy from some other form of energy. It is gnerally termed as a power station.

Generator-field Control : Refers to the method of controlling an electric motor, like that powering an electric lift, by the use of a motorgenerator. A direct voltage applied to the armature of the lift motor gets varied by changing the strength and polarity of the generator field. This is also termed as variable-voltage control.

Generator : Refers to a machine which is used for converting mechanical energy into electric energy. Its design has been based on the fundamental laws of electromagnetic induction (Fig. 4).

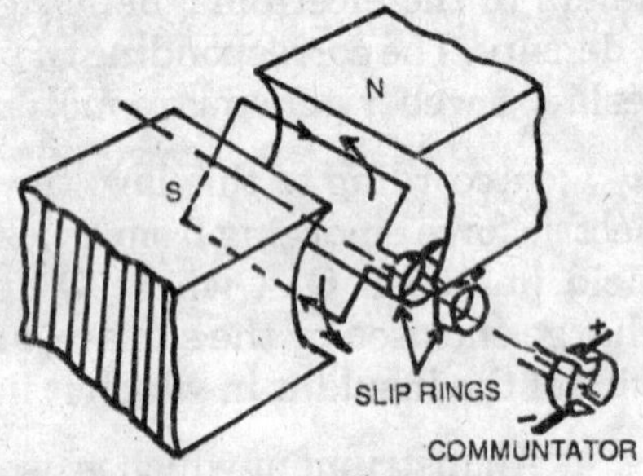

Fig. 4. Basic a.c. and d.c. generators. The slip rings will provide an alternating supply. The commutator will provide a unidirectional output.

Geostationary Relay Station : A space vehicle which moves in space at a fixed distance of some 36000 km from the surface of the Earth and in synchronism with the Earth's durial rotation so that it is remaining stationary above a given spot on the Earth's surface. It has been equipped to receive and re-transmit radio signals from a transmitting station situated on the earth. The re-transmission may occur either narrowly directional or in the form of a wide beam to a multitude of receiving stations on the Earth.

Geothermal Power Station : Refers to a power station that uses the internal heat of the Earth as the source of energy.

German Silver : A resistance alloy having copper, zinc and nickel.

Germanium : A tetravalent metallic element which is a semiconductor. It has low conductivity at room temperature and increasing conductivity with rising temperature. When it is allowed with very small but accurately controlled proportions of trivalent or pentavelent metallic impurities, thereby producing p-type or n-type germanium respectively, it finds use in the manufacture of crystal diodes and transistors. Germanium is having a melting point around 950°C.

Germanium Rectifier : A junction diode having p-type and n-type germanium for use a rectifier. It will be seen from the typical characteristic curve (Fig. 5 that the reverse current gets greatly increased at higher temperatures. This limits the voltage per cell that can be used in practical applications.

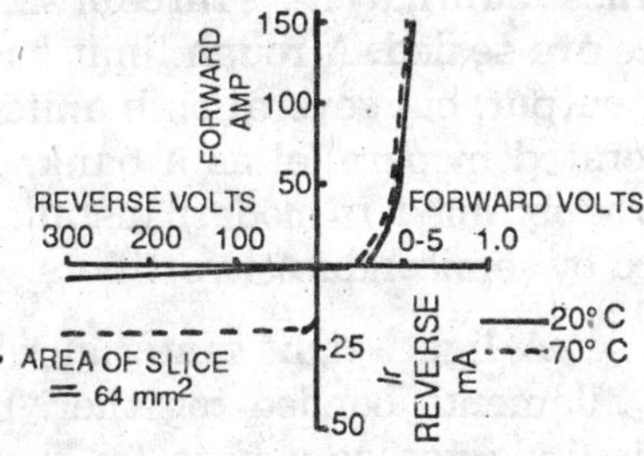

Fig. 5. D.C. Charateristic of germanium junction rectifier.

Getter : The term used for chemically active metal such as barium or magnesium which is used to increase the vacuum in electronic valves, lamps etc. After the component is sealed off from the vacuum pump, the getter material inside it gets volatilised by heating it electrically. During volatilisation, the getter combines with the residual gases in the vessel so that they get fixed as chemical compounds deposited on the walls.

Giga : A decimal prefix which indicates multiplication by 10^9 (abbreviation: G).

Gilbert : Refers to the electromagnetic c.g.s. unit of magnetomotive force. One gilbert has been equal to $10/\pi$ ampere-turn.

Giorgi System : System of units which is adopted internationally in a rationalised form as the m.k.s. system. Giorgi reported in 1901 that the confusion of electrical unit system (*i.e.*, the co-existence of electromagnetic and electrostatic c.g.s. sytems, together with the 'pratical' system) could be resolved by using the metre and kilogram in place of the centimetre and a gram as units of length and mass respectively. Such a system would incorporate the practical units and extend them. Now this system has been superseded by SI.

Glass : hard, amorphous brittle substance which finds used in electrical engineering for insulating envelopes and line insulators. Its melting point lies between 800° and 950°C; the sensitivity of soda-lime glass has been $5\times10^{11}\Omega$ cm and of Pyrex glass, $10^{14}\Omega$ cm.

Glass-bulb Rectifier : Refers to a form of mercury-arc rectifier whose bulb may have three or six arms into which the anodes are sealed. A rough limit has been 500 A at 600 V.d.c. output, but several bulb units have been commonly operated in parallel as a bank, enabling higher outputs to be obtained. In modern installations it has been superseded by semiconductor rectifiers.

Glass Fibre : A lightweight material which is made of fine glass filaments bonded together. In the electrical industry it finds use as a base for flexible insulation, sleevings, winding wires and cables.

Glow Conduction : Refers to a silent form of gas conduction which in part of its path has been luminous.

Glow-conduction Lamp : A tube having neon, argon or krypton inert gas and provided with two electrodes.

Graded-gap Machine : Refers to an machine in which the distance between the armature surface and the pole face gets varied from point to point along the pole arc in such a way as to produce at no-load an unsymetrical distribution of flux. The maximum value has been on that side of the centre line of the pole opposite to which it tends to get moved by the action of the armature ampere-turns on load; under full-load conditions, therefore, these two effects substantially are able to counteract each other.

Grading Shield : Refers to a concentric circular conductor which is designed to improve the voltage distribution along a string of insulators.

Grain-oriented Steel : Refers to a silicon steel in which the individual crystals get aligned by suitable treatment to provide strongly marked directional magnetic properties. Sheets of this steel are having higher permeability and lower iron loss in the direction of the grain than untreated sheets.

Gramme Winding : Alternative name for ring winding.

Graphic Instrument : An instrument which is used for producing a graphic record of the quantity measured. It is having a pointer in the form of pen which is moving over a paper chart. It is also known as a chart-recording instrument, a grapher, a recorder or a recording instrument.

Graphite : An allotropic form of carbon. Graphite brushes for collecting the current from the commutator of an electric machine are having higher conductivity and better lubricating properties than ordinary carbon brushes. Very pure graphite has been used for making anodes in thermionic tubes.

Graphite Brush : See graphite.

Grassot Fluxmeter : A fluxmeter having a sector-shaped horizontal-scale pointer instrument.

Grenz Rays : Usually called infra-roentgen rays. These are produced at tube voltages of 5-20 kV and find use for treatment of superficial lesions (from the German word 'Grenze' meaning 'boundary').

Grid :

(1) A cast or stamped unit which forms part of a resistor.

(2) The framework in a lead-acid accumulator which is supporting the active material of a pasted plate.

(3) An electrode in a gas-filled tube which is used for controlling the current flow between two other electrodes.

Grid System : Refers to an electric power transmission network which is connecting together, within a given area, all the sources of power and all the points of bulk supply to consumers.

Growler : A device which is used for detecting short-circuits in a winding coil. It comprises of a stack of stampings wound to suit the supply voltage, and acts as a transformer with no secondary winding. It finds use to induce a flux in the component under test. A short in a coil between the poles of the growler makes a distinct pull on a light steel feeler held over the slot.

Guard Wire : In overhead-line construction, it refers to an earthed conductor which has been arranged to protect physically the line conductors. It may run beneath the line conductors to disallow them from falling to the ground, or above them to disallow other conductors from falling on them.

Gyro-machine System : Refers to an electric method of driving vehicles by using the energy stored in a flywheel. A gyrobus is having a large, heavy flywheel which is mounted horizontally under the floor. Charging columns at the roadside at less than six-mile intervals make a three-phase supply to get connected for two or three minutes to a squirrel-cage induction motor that accelerates the flywheel to over 3000 rev/min. The motor then is able to operate as a self-excited generator, driving traction motors on the vehicle.

Half-cell Half Element : Of an electrolytic cell, this term refers to one electrode and the portion of the electrolyte in contact with it. It is also termed as a single-electrode system.

Half-power Point : Refers to the point on a response characteristic which represents half the power quantity of that of the maximum point. It is also termed as a 3 dB point.

Half-value Layer : Refers to a measure of the quality of a heterogeneous X-ray beam which may be defined as the thickness of a specified material that reduces the intensity of the beam to 50% of its original value.

Half-wave : A term which finds use especially in connection with rectifiers to describe a system in which one half-cycle of an alternating voltage generates conduction current, while the other half-cycle produces a comparatively negligible or no current.

Hall Angle : Refers to an angle through which equipotential surfaces in a conductor get tilted due to the influence of mutually normal electric and magnetic fields.

Hall Effect : The term used for the combined action of mutually normal electric and magnetic fields on the flow of conductiot electrons in a conductor. For electron flow from left to right along a thin strip and for the magnetic field acting vertically upwards, electrons will get deflected towards the from edge and three would be set up between opposite points on the edges potential difference called the Hall voltage. In effect, the equipotentials get tilted through an angle known as the Hall angle.

Hall Voltage : Refers to the potential difference bet- ween opposite edges of a conductor which is caused by the influence of mutually normal electric and magnetic fields.

Hanger :

(1) Refers to a field which is used for separating and insulating the overhead contract wire of a traction system from a transverse wire or structure.

(2) Refers to an insulating plate which is standing on edge in an accumulator and supporting the accumulator plates by means of lugs.

Harmonic Distortion : Refers to a change in waveform, due to the inclusion of additional frequency components.

Harmonic Factor : Refers to the ratio of the r.m.s. value of all the harmonics of a non-sinusoidal quantity to the r.m.s. value of that quantity.

Harmonic Response : Refers to the ratio output/input of a dynamic system (such as an automatic control system servomechanism or electric-machine) for an input signal of sinusoidal waveform and specified frequency.

Hay Bridge : It is a real-product a.c. bridge of the form shown in Fig. 1. Usually the bridge gets balanced by adjustment of the resistor in series with the capacitor, and

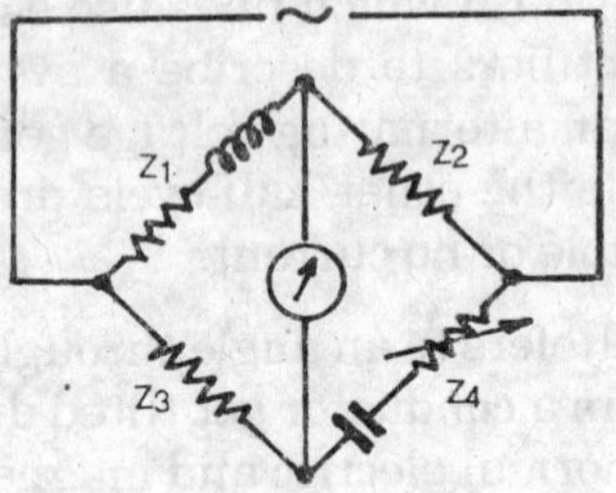

Fig. 1. Basic circuit of a Hay bridge

of one of the non-reactive arms. The balance is dependent upon the frequency.

H-class Insulation : Refers to one of seven classes of insulating materials for electric machinery and apparatus on the basis of thermal stability in service. Class H insulation has been assigned a temperature of 180°C.

Heat Pump : Refers to a device which is used for transforming low-grade heat (as from the air, rivers, sea or soil) into higher. Heat gets extracted by a refrigerant, which then gets compressed and transfers its heat to air or water that can be used for space or process heating.

Heat Run : Test of an electric device like a machine or transformer under operating conditions with full load current passing in the windings to determine the temperature rise. With two machines the test has been frequently carried out as in the Hopkinson test.

Heat Sink : Refers to a component in direct contact having a device and with a sufficient thermal capacity to disallow excessive temperature rise of the device during its normal periodic operation.

Heating Inductor : Refers to a primary work circuit which is used to induce high-frequency currents for the induction heating of conductive material.

Heating Resistor : The term used for wire or other resistance material which is used as the source of heat in an electric heating element.

Heaviside-campbell Bridge : Refers to an a.c. bridge that introduces mutual inductance, the primary winding

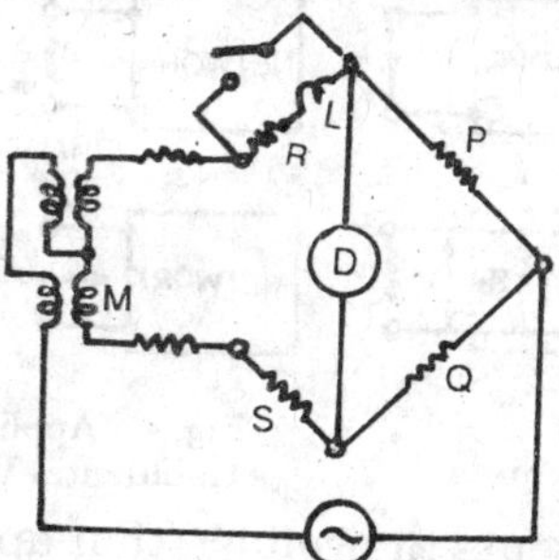

Fig. 2. Basic circuit of a Heaviside—Campbell bridge.

of the mutual inductor being connected in the supply lead. Two balances have been obtained, without and with a test coil RL in circuit. Then with P=Q, the values of the known would be

$$R\text{-}S_1\text{-}S_2 \text{ and } L=2(M_2-M_1) \text{ (Fig. 2).}$$

Helical Winding : Refers to a transformer winding which consists of a single coil of one or more layers with its axial length usually being considerably greater than its diameter.

Helmholtz - Norton Theorem : Refers to the dual of the Helmholtz - Thevenin theorem. The voltage across any branch of admittance Y in a network that is having several current sources is given as follows : $V=1_0/(Y+Y_0)$, where I_o represents the current that flows in the branch when short-circuited, and Y_0 represents the admittance of the network viewed from the branch terminals. The theorem in effect is able to convert the network (apart from the branch Y) into an equivalent current generator of current I0 and shunt admittance Y_0. The applicability of the theorem has been the same as that specified for the Helmholtz-Thevenin theorem (see below). The stages in the application are shown in Fig. 3. When Y_0 is found, all current sources get replaced by their internal shunt admittances.

Helmholtz-Thevenin Theorem : Refers to a particularly useful method of network solution. If states that the current in any impedance Z, forming a branch of a network having one or more sources of e.m.f., is $I=E_0/(Z+Z_0)$,

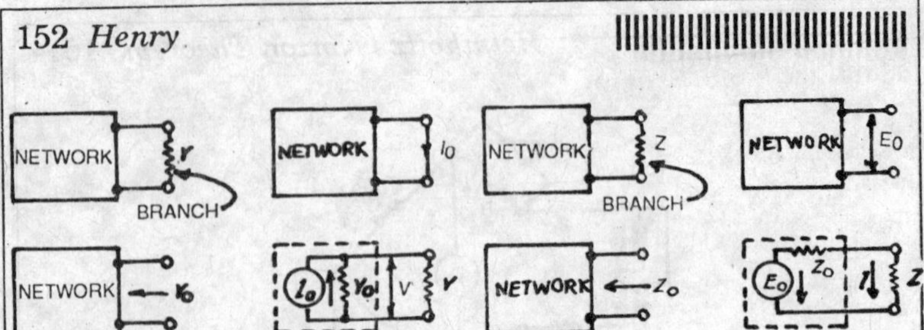

Fig. 3. Application of Helmholtz–Norton theorem.

Fig. 4. Application of Helmholtz–Thevenin theorem.

where E_0 represents the voltage that appears across the branch when open-circuited, and Z_0 represents the impedance of the network looking in at the terminals of the branch, as depicted in Fig. 4.

Henry : It is the SI unit of inductance (symbol : H). For a current of 1 A, an inductor of 1 H will be having a total magnetic-field energy storage of $\frac{1}{2}$ J, and the flux-linkage will be 1 Wb-turn. For a rate of change of current of 1 A/s the voltage induced in an inductor of 1 H would be 1 V.

Heroult arc Furnace : It is an indirect-arc furnace. Direct current was originally used, but modern equipment invariably uses three electrodes connected to a three-phase supply (see Fig. 5). Power has been supplied by a step-down transformer with variable voltage to suit the requirements of the smelting or refining operation. Currents range up to 10 kA, the lower values being used for the refining stages. Separate control has been normally applied to the three electrodes, so that they may be operated individually.

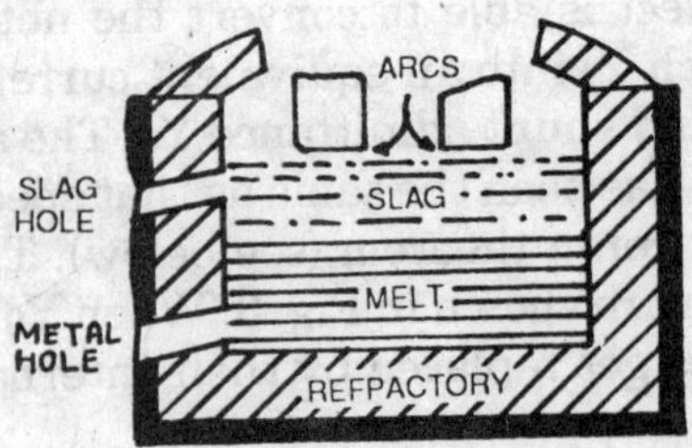

Fig. 5. Principle of the Heroult arc furnace as used for steel making.

Hertz : It is the SI unit of frequency (symbol : Hz). 1 Hz=1 period per second.

Heteropolar Machine : Refers to an electromagnetic machine in which the conductors pass successively through magnetic fields of opposite sense.

Hexagon Voltage : Refers to the voltage between any two consecutive lines of a symmetrical six-phase system. Also termed as line voltgage or mesh voltage.

High Energy Rate Forming : The term used for the forming of metals by using high-intensity shock waves which are usually produced in water and develop pressures approaching 7000 MN/m^2. The energy source could be the discharge of a high-voltage capacitor, the magnetic field of a coil placed parallel to the workpiece, or an explosive. The magnetic technique is regarded the most promising, but the spark machine with a capacitor is cheaper.

High-frequency Heating : Refers to the production of a temperature rise in a material by making high-frequency currents to flow in the material.

High-frequency Induction Furnace : Refers to an induction furnace which is using the fact that at high-frequency current at high voltage passing through a water-cooled coil of copper tubing gets induced eddy currents in the metal within the coil (see Fig. 6). It largely finds use for the production of high-grade steel and high-melting-point alloys. It is sometimes referred to as a coreless induction furnace.

High-frequency Power Generator : A device which is used to produce power up to about 100 kW for high-frequency heating. It may be electronic, or may comprise a

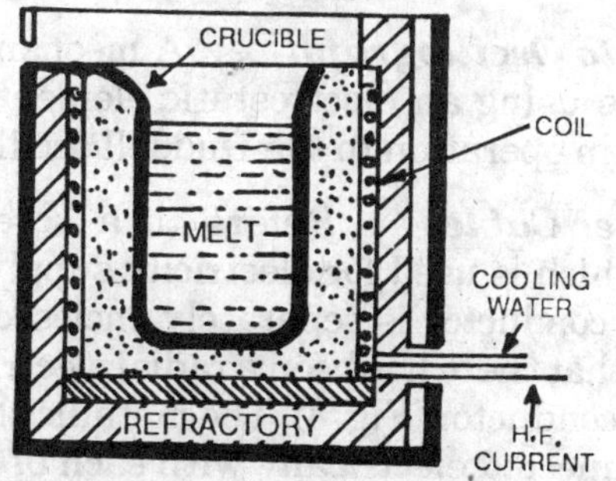

Fig. 6. High-frequency, or coreless, induction furnace.

high-speed rotating machine. The latter is commonly termed as an inductor alternator.

High-rupturing-capacity Fuse : Refers to a high-performance cartridge fuse. It is also known as a high-breaking-capacity fuse (h.r.c. or h.b.c.). Its minimum repturing capacity has been normally in excess of 16500 A. It is designed to limit current to a safe value (*i.e.*'cut-off') and to act rapidly.

High Voltage : A term which, when normally implies a voltage exceeding 650 V. Abbreviation : h.v.

Hilborn Loop Test : A modified loop test which is similar to the Fisher loop test. It has been of value when two sound conductors of cross-section different from that of the faulty conductor are to be used. In Fig. 7 at balance, x, the distance of the fault along the cable would be given by

$$x = L\frac{b}{a+b+r}$$

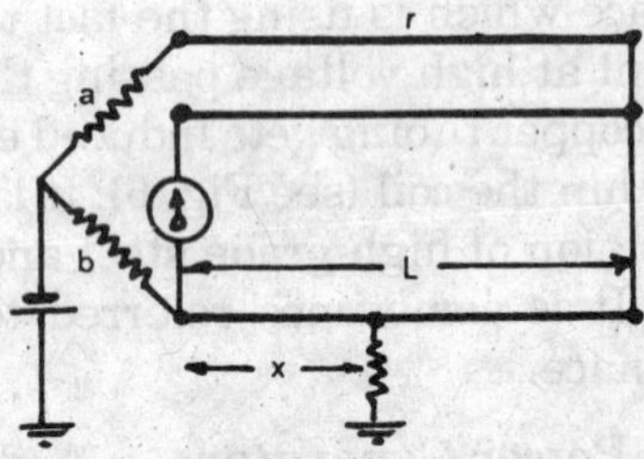

Fig. 7. Hilborn loop test for a faulty conductor.

Resistance r may be neglected if small compared with a+b.

Ho and Koto Oscillograph : A mechanical oscillograph which is using an electrostatic element. It is otherwise similar in operation to the Duddell oscillograph.

Hochstadter Cable : Refers to a screened multicore cable which is used for electricity supply. The insulation of each conductor is separately enclosed in a conductive film so that there has been a radial electric field surrounding the conductor (Fig. 8). The matallic sheath of the cable gets connected electrically with each of the films. In this way, tangential stresses get eliminated, so enabling the

Fig. 8. Hochstadter multicore screened cable.

design to be used up to 50 kV.

Hole : The term used for the concept of a positive charge carrier in conduction by a p-type semiconductor. It denotes the deficiency of a valency electron in an atom of acceptor impurity.

Hollow Conductor : Refers to a tubular conductor which finds use

(a) for large-current high-frequency conductors where the skin effect would make any interior regions inoperative,

(b) in large turbo-generators to facilitate direct cooling by air, hydrogen or water, and

(c) on high-voltage transmission lines where the diameter of the conductor needed for elimination of corona loss has been significantly greater than that needed for current conduction or mechanical strength.

Home Office Switch : Refers to a switch in which all the metal parts have been covered by insulating material or earthed, to guard against the possibility of electric shock.

Homopolar Component : Alternative term for zero phase sequence.

Homopolar Machine : Refers to an electromagnetic machine in which there has been only one active pole, compared with the two (or multiple of two) in the common heteropolar machine. In principle the machine is having a conductor so disposed in a magnetic field that, in motion, it cuts through the magnetic field continuously in one direction.

Hopkinson Tarrif : Obsolete term which is superseded by two-art tariff.

Hopkinson Test : Refers to a back-to-back test for two similar machines. Figure 9 depicts the connections for testing two d.c. machines. These machines get coupled mechanically, and joined electrically in parallel to a.d.c.

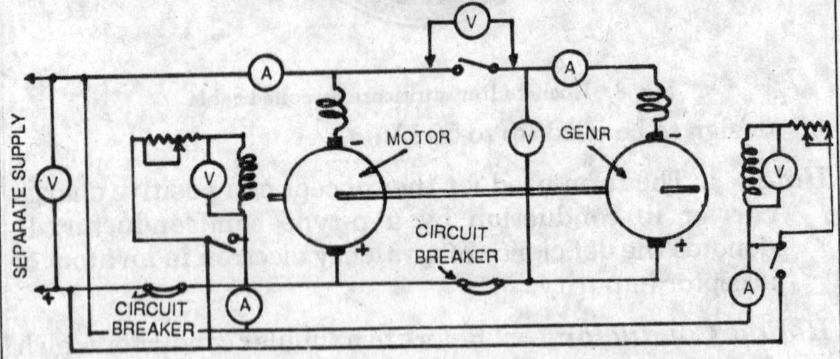

Fig. 9. Diagram of conr ection for Hopkinson test on two d.c. machines.

supply which compensates for the losses. One machine is acting as a motor to drive the other as a generator with equal currents in the two machines.

Horizontal Plugging : Refers to a method of switch- gear busbar selection in which the connections have been carried on a framework that moves horizontally.

Horn Gap : Refers to a rod gap having extended electrodes up which the arc may travel owing to its heat, increasing its length so that it becomes self-extinguishing.

Horsepower : A practical unit of power which is thus equal to about 746 W (J/s).

Hot-cathode Rectifier : Type of mercury-arc rectifier.

Hot-wire Instrument : Indicating instrument which consists of a fine wire that carries the current to be measured, the expansion produced being mechanically amplified by permitting the sag to operate a spring-controlled pointer. It is having a square-law scale. True r.m.s. value has been measured on alternating current; it also reads direct current.

H.R.C. : High-rupturing-capacity Fuse.

H-type Cable : Hochstadter cable.

Hunter : A form of synchro which is having three-phase windings on both stator and rotor. It finds use in servo systems especially for operating hydraulic valves and indicating differences in angles.

Hunting : Refers to an oscillatory phenomenon with particular reference to synchronous machines. Under steady load, the driving and load torques of a synchronous machine would balance for a given load angle. Any sudden change of condition will set up a transient oscillation superimposed on the steady mean speed. The unbalanced part of the load angle is releasing an excess torque to act on a mechanical oscillatory system formed by the machine, producing the possibility of hunting (or phase-swinging).

H.V. : High voltage.

H.V.D.C. Power Transmission : Refers to a scheme to transmit large quantities of electric energy over a high-voltage d.c. system which gets inserted between two converting stations and coupling two a.c. systems. Each of the two station has been equipped with a high-voltage static converter, which operates either as rectifier or inverter depending on the direction of energy flow. The scheme is has been used for energy transport over very long overhead lines or over underground or submarine cables, or has been used to couple two asynchronous a.c. systems. High-power grid-controlled mercury-arc valves are used in the converters, and high-power triode thyristor valves are now coming into use.

Hybrid Coil : Refers to a type of transformer which is used for reducing side-tone in telephone subscribers, sets, and also in telephone lines.

Hydro-alternator : Refers to an alternator which finds use in conjunction with a water turbine in hydroelectric generating plant. As a water turbine is a low-speed machine compared with a steam turbine, the hydro-alternator has been much larger and heavier than one of equivalent capacity for use in a steam generating station. it is frequently convenient to build the units with vertical shafts.

Hydroelectric Generating Plant : Refers to a generating equipment which is driven by water turbines, the water being supplied from a reservoir (possibly pumped-storage) or tidal barrage.

Hydrogen Cooling : Refers to a method of cooling large electric machines of the turbo type which can readily get enclosed, Hydrogen, compared with air, is having 1/14 of the density, reducing windage loss and noise; 14 times the specific heat; $1\frac{1}{2}$ times the heat-transfer capability, so more readily abosrbing and giving up heat; and 7 times the thermal conductivity, so reducing temperature-gradient. It is also having corona-reducing properties, and will not support combustion so long as the proportion of hydrogen to air in a mixture exceeds 3 1. As a result, hydrogen cooling at 100, 200 and 300 kN/m^2 (1, 2 and 3 atmospheres) can be able to raise the rating of a machine by 15, 30 and 40% respectively.

Hysteresis : Refers to a phenomenon in which the effect due to an impressed stimulus has been two-valued; a lower value when the magnitude of the stimulus is rising, and a higher one when the stimulus has a falling magnitude. The effect is seen in gears ('backlash'), springs, ferroelectric and ferromagnetic materials.

Hysteresis Loop : Refers to a closed loop in a B/H curve or a D/E curve, which is formed when a ferromagnetic or a ferroelectric material is taken through a complete cycle of magnestisation or electric polarisation respectively.

Hysteresis Loss : Refers to the dissipation within a ferromagnetic or ferroelectric material taken through a hysteresis loop. Its magnitude has been proportional to the area of the hysteresis loop.

I a.c.s. : International annealed copper standard.

Ideal Source of Electric Energy : Refers to a theoretical device in the analysis of electric circuits based on equivalent generator.

Ideal Transformer : Refers to a transformer which is devoid of loss in windings and core, with perfect coupling between primary and secondary windings, and with a magnetic circuit of zero reluctance. It finds use in the theoretical analysis of magnetically coupled electric circuits.

Idle Component : Alternative name for reactive component.

Ignition : Refers to the intentionally caused process to initiate gas conduction in a gaseous medium.

Ignition Coil : Induction coil used with internal-combustion engines to convert the l.t. current supplied by the battery into the h.t. current needed by the sparking plugs.

Ignition Voltage : The voltage which is needed to cause ignition.

Ignitron : Refers to a single-anode, pool-cathode mercury-arc rectifier. The ignitron is different from the other mercury-arc rectifiers (single and multi-anode) in the method which is used for ignition of the arc. As in the case of the thyratron, the arc, once started, cannot normally get extinguished except by interrupting the anode supply or by the anode-to-cathode voltage becoming less than the arc drop (as would happen every cycle with an a.c. supply). When ignitrons find use in a rectifier circuit, controlling the phase position of the igniter pulse relative to the positive half-cycle of the a.c. supply to the anode gets varied the conduction period as needed. It has een analogous to the grid control or a thyratron.

Ilgner System : Refers to a method of smoothing out the peak demands on a motor/generator system which is used in conjunction with a Ward-Leonard control. A heavy flywheel gets mounted on the d.c. generator shaft and arranged to give up some of its stored energy during the peak-load period.

Illuminance : Refers to the quotient of the luminous flux which is incident on a surface area by that area. The SI unit is the lux (symbol: lx).

Image Impedance : Refers to impedance which, when connected to the terminals of a network, becomes equal to the impedance presented to it. There thus occur no reflection loses at the junction.

Immersible Switchgear : A switchgear which is capable of working indefinitely without detriment while submerged to a considerable depth in water.

Immersion Heater : Refers to a heater, in which the heat has been supplied by a metallic resistor completely insulated from the water. It gets fitted into the hot-water supply tank.

Immittance : A term which is used in the theory of network analysis to denote either impedance or admittance.

Impact Grinding : Refers to a form of machining in which a tool of the necessary profile has been vibrated longitudinally and cuts the work by means of an abrasive. If the frequency has been ultrasonic, using a magnetostrictive transducer driving a drill that vibrates longitudinally through a solid stub, the work could be drilled with an abrasive by a microscopic chipping process. A frequency of about 20kHz is used, the stub being tunned mechanically to vibrate at the designed frequency. The amplitude of vibration has been of the order of 0.025 mm.

Impedance : Refers to the complexor Z providing the quotient of the r.m.s. voltage phasor V by the r.m.s. current phasor I in a circuit with sinusoidal voltage and current. It has been as follows

$$Z=V/I=Z\phi$$

where ϕ denotes the angle by which the voltage leads the current. Then $Z \cos \phi = R$, the resistance, and $Z \sin \phi = X$ the reactance.

Impedance Bond : Refers to an inductive form of continuity bond which is used at the extremities of a railway track circuit so that d.c. traction current will pass through it conductively unhindered, where as the a.c. track-circuit currents have been presented with a high reactive impedance.

Impedance Drop : Refers to decrease in output voltage of circuit due to the internal impedance voltage within the circuit.

Impedance Triangle : Refers to a phasor triangle which is formed by R, X and Z (rdsistance, reactance and impedance) as in Fig. 1. Using complex algebra, the impedance complexor may be as follows:

$$Z=R+jX$$

The modulus $Z=\sqrt{(R^2+X^2)}$

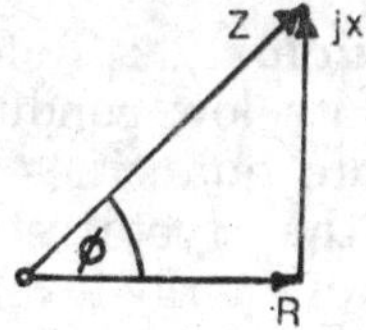

Fig. 1. Impedance Triangle.

Impedance Voltage : Refers to the voltage due to a current which is flowing through the impedance of a circuit. it is the phasor sum of the resistance voltage and the reactance voltage.

Impregnated Carbon : The carbon rod which is used as an electrode in a carbon arc lamp. The carbon is mixed with other materials to provide a required colour to the arc.

Impregnation : Of an insulating material, this term refers to the substitution of the air between its fibres by a suitable substance (such as a wax or varnish).

Impulse Circuit-breaker : Oil circuit-breaker, which is having a low oil content, in which an oil blast to extinguish the arc has been produced mechanically by using a spring or compressed-air operated piston.

Impulse Clock System : Refers to a self-contained clock system which gets characterised by the fact that the minute hands of the salve clocks move round in half-minute steps, under the control of a master-clock.

Impulse Generator : Deprecated term for surge generator.

Impulse Ratio : Deprecated term for surge ratio.

Impulse Test : Deprecated term for surge test.

Impulse Turbine :

1. Refers to a steam turbine in which velocity compounding is used with each stage of expansion, associated with two or more rows of moving blades separated by fixed guides to redirect the steam into the next row of moving blades without stock but with unaltered velocity.
2. Also, refers to a water turbine such as the Pelton wheel.

Impulse Voltage, Current : Deprecated terms for surge voltage, current.

Impurity Semiconductor : Refers to a semiconductor which is having its low conductivity prodived by the presence of minute quantities or foreign atoms or by deformations in the crystal structure. The impurities 'donate' electrons of energy-level that can get raised into a conduction band (n-type) or they can attract an electron from a filled band to leave a hole, or electron deficiency, the movement of which is corresponding to the movement of a positive charge (p-type).

In-phase Component : Alternative name for active component.

In-quadrature Component : Alternative term for reactive component.

Io-situ-Conduit : Refers to the passages which are formed in concrete foundations by means of inflatable pneumatic tubing. When the concrete is set, the tube get deflated and withdrawn, as fish wire' being pulled in at the same time.

Inclined Catenary Suspension : In electric traction, this term refers to a form of consutruction for overhead conductors when the supoort structures have been on alternate sides of the track. The droppers attaching the contact wire to the support catenary have been thus necessarily inclined to the vertical.

Incremental Inductance, Permeability, Resistance : Re fers to a measure of the inductance, etc., in a ferromagnetic core inductor, or other component which is having non-linear relationships between the appropriate properties.

Independent Feeder : A feeder which is used solely for supply to a substation or a feeding point, and not as a trunk feeder. It is also called a dead-ended feeder or radial feeder.

Independent-trip Circuit-breaker, Starter : Refers to a circuit-breaker or starter which is tripped with the help of an auxiliary power supply, regardless of the state of the main circuit in which the breaker is connected.

Indicating Instrument : A measuring instrument in which the value of the quantity being measured is revealed by the relative positions of a pointer and scale, or a similar means.

Indirect-arc Furnace : Refers to an arc furnace which is having a cylindrical or spherical body with axially disposed horizontal elecrtrodes between which the arc has been struck. The furnaces are single-phase, and so special equipment is usually required to take a balanced load. Malleable iron, alloy steels non-ferrous alloy, etc., have been melted in this type of furnace.

Indirect Stroke : Refers to a stroke of lightning which, while not striking a transmission system, influence it by inducing a voltage in it.

Induced-draught Ventilation : Refers to a form of machine ventilation in which the ventilating air has been drawn through the machine by some means external to the machine itself.

Induced e.m.f. : See induced voltage.

Induced Moving-magnet Instrument : An instrument hav ing a movable piece of magnetic material which gets deflected by the resultant field produced by a fixed current-carrying coil and a permanent magnet at an angle to the coil.

Induced-overvoltage Test : Refers to an overvoltage test in which a transformer gets overexcited at an enhanced supply frequency, to a voltage about 67% greater than the nominal system voltage.

Induced Voltage : Refers to the voltage which is produced in a circuit by electromagnetic induction.

Inductance : Refers to the ration of the magnetic flux-linkage of an electric circuit and a current. The SI unit is the henry (symbol: H)

Induction Coil : An open-cored transformer which is used for developing damped high-voltage transients from a d.c. primary supply by using an interrupter. A descendant in common use is the ignition coil used in petrol engine.

Induction Furnace : An electric furnace for melting metals in which the heat is produced in the charge itself. There are two types, the high-frequency induction furnace and the low-frequency induction furnace.

Induction Generator : Refers to a generator which is similar in construction and operation to an induction motor. An induction motor runs at a speed slightly below synchronism and draws power from the supply to drive its load.

Induction Heating : Refers to a form of high-frequency heating which is applied to conductive materials. When elecrtically conductive material is kept in an alternating magnetic field, the voltage induced by transformer action makes alternating currents to flow in the conductor. These corrents cause resistance losses that produce heat in the conductor.

Induction Instrument : Refers to an indicating instrument in which two magnetic fields have been produced in one or two adjacent iron circuits excited by current, the

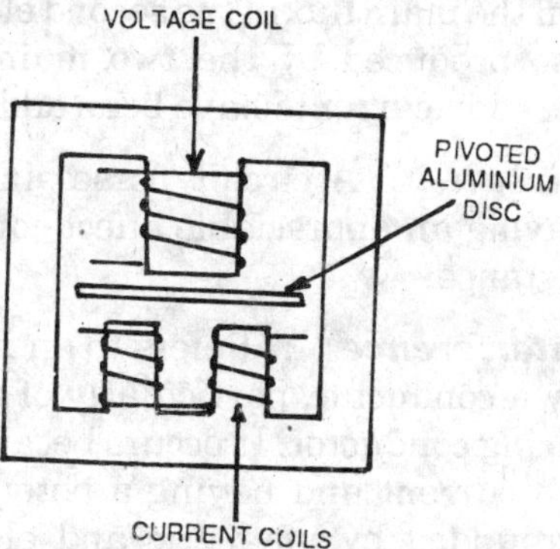

Fig. 2. **Induction wattmeter. An induction instrument has the advantage of a long-scale arc.**

two magnetic fluxes possessing a small phase-angle between them to produce initial movement and so to indicate on a fixed scale. Only alternating current of power frequencies could be measured, but as a wattmeter (see Fig. 2) the type has been very useful, as the scale arc may be about 300° and so be readily readable from a distance.

Induction Motor : Refers to an a.c. motor in which the current in the stator winding gives rise to a rotating flux which induces a current in the winding of the rotor, thereby producing the necessary torque.

Induction Pump : Refers to a type of electromagnetic pump in which the current in the liquid is induced by the field. Induction pumps have been restricted to the handling of alkali metals.

Induction Regulator : Refers to an electromagnetic device which is having a primary winding in parallel, and a secondary winding in series, with an a.c. supply. The voltage impressed on a load could be adjusted by changing the relative positions of the primary and secondary windings.

Induction Relay : Refers to a relay which is using either a rotating disc or a rotating cup to operate the contacts. In the former case a metal disc gets mounted so that it has been free to rotate between the poles of two electromagnets. Torque is produced by the interaction of eddy currents, produced in the disc by one magnet as-

sembly, on the main flux of the second electromagnet. The torques so produced by the two main fluxes with the respective eddy currents have been additive.

Inductive Circuit : A circuit possessing self-inductance that is having an appreciable effect compared with that of its resistance.

Inductive Interference : Refers to an induction which is caused by a conductor, particularly of a power line, in a neighbouring conductor. It occurs because any conductor carrying a current and having a potential to earth has been surrounded by magnetic and electric fields. The voltage magnetically induced in the neighbouring parallel conductor is dependent upon the strength of the magnetic field linking the two conductors, the frequency, the resistivity of the earth, the length of the parallelism and the distance between the lines.

Inductive Load : See lagging load.

Inductive Potentiometer : Refers to a precision-wound toroidal auto-transformer having two sliding contacts. In its basic use the auxiliary contact has been set for initial calibration, while the main contact has been moved in accordance with some rotary mechanical input, to produce a corresponding voltage to an accuracy better than ± 0.1%. Fig. 3 depicts the essential features of the 'I-pot'. It comprises in effect a low-loss variable auto-transformer of small leakage (and therefore, low internal impedance). The input voltage could be divided accurately by the main sliding contact.

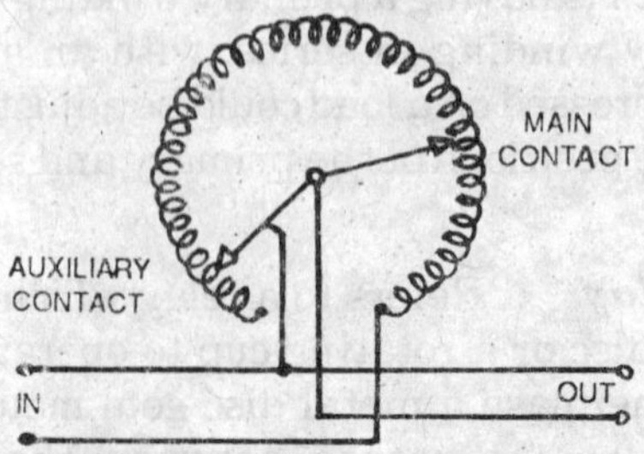

Fig. 3. Inductive Potentiometer. Combinations of 1-pots can be arranged to produce outputs that are fractions, multiples, and other functions, of the input.

Inductor : Refers to a device whose primary characteristic has been the property of inductance. Its form has been almost invariably a compact coil.

Inductor Alternator : Refers to a rotating high-frequency generator which is usually driven by a standard induction motor running at 3000 rev/min.

Industrial Capacitor : A capacitor which is distinguished by its large size, both physically and electrically. Two main forms of construction are used known as unit capacitors and tank capacitors. Tank type capacitors are normally large-up to at least 500 kV Ar. In all these capacitors the dielectric is liquid-impregnated paper, the liquid being either a refined mineral oil or a synthetic chemical having suitable electric properties.

Inert Cell : A closed dry primary cell which is having solid ingredients that form an electrolyte only when water is added.

Inertia Constant : Refers to a measure of flywheel effect. The inertia of the rotating parts of an electric machine (with its prime mover or mechanical load) influences its stored kinetic energy when running. A large and heavy machine will therefore take sufficient a time to start and stop. The requisite flywheel effect could be expressed in terms of an inertia constant H, defined as

$$H = \frac{Storedenergy, joules}{rating,\ voltampere}$$

The values of H normally needed have been between 2 and 9 seconds.

Infinite Busbar : Refers to a busbar system of rigidly constant voltage and frequency. The behaviour of a single synchronous or asynchronous machine connected to a supply system fed by an extensive network of generators may, to an approximation, be regarded as working on a infinite busbar system. Then no change of load will alter the terminal conditions of voltage or frequency. It is not possible to achieve the conditions of an infinite busbar in practice except under rare, rigidly constant conditions..

Influence Machine : Alternative name for electrostatic generator.

Infra-red Heating : Refers to production of heat by using infra-red rays. These rays may be produced electrically by either using a 'dull-emitter' resistance element operating at black heat, or a special infra-red tungsten filament lamp, designed to produce 'light' at a wavelength within the infra-red band; such lamps are generally of 250 W rating.

Inherent Regulation : Refers to the change in output terminal voltage of a generator or converter when the load gets reduced from the full rated value to zero, all other conditions remaining constant. It has been usually expressed as a percentage of the normally value.

Inhibited Oil : Refers to the transformer or switch oil, the deterioration of which during its working life gets retarded by the use of anti-oxidants, particularly oxidation inhibitors.

Instantaneous Value : Refers to the magnitude, at any given instant in time, of a time-varying quantity like voltage, currents, charge, flux, etc.

Instrument Transformer : Refers to a transformer which is used to reproduce the voltage and current values of a circuit, generally a power or test circuit, in true proportion and phase relationship but of dimensions suitable for application to measuring or protective instruments.

Insulance : Obsolete term for insulation resistance.

Insulated system : Refers to a distribution system that is having no point connected to earth.

Insulated-return System : Refers to a supply system for electric traction in which both the outgoing and the return conductors have been insulated from earth.

Insulating Material : A material which is offering high resistance to the passage of electric current. Insulating materials have been of all forms: gaseous, liquid and solid; organic and inorganic; natural, refined, manufactured or synthesised. An insulating material is also known as an insulant. A dielectric is a particular type of insulant.

Insulating Oil : Refers to an oil which is used as an insulating medium in transformers and switchgear. It is a pure hydro-carbon mineral oil, clean and free from matter and without additives.

Insulation Classification : The term used for the classification of insulating materials in terms of their operating temperature limitations.

Insulating Grading : Refers to the adjustment of an electric field in insulating materials so as to secure a better utilisation of the properties of the materials, especially as regards their electric strength.

Insulation Leval : Refers to a design figure which is used to indicate the degree of insulation of a device or power transmission system. It signifies the voltage at which the insulation is tested.

Insulation Resistance, Conductance : Refers to the resistance or conductance which is offered by an insulating material separating two conductors having a potential difference, or a live conductor and earth.

Insulator : A component which is made of insulating material. It used to separate a conductor from earth or from another conductor.

Integrated Circuit : Refers to a combination of a number of circuit elements which have been inseparably interconnected in such a manner that a single indivisible unit for external connection in a circuit has been formed.

Interating Frequency Meter : Alternative name for master frequency meter.

Integrating Meter : Refers to a measuring instrument which adds up the value of the quantity measured with reference to time. A single-phase meter for metering power supplied is having a voltage system consisting of E-shaped iron laminations energised by a multi-turn voltage coil, mounted above a second set of U-shaped laminations energised by a current coil carrying the load current. Between the two systems has been an air-gap in which a flat circular disk mounted on a vertical spidle rotates in special bearings.

Integrating Wattmeter : Alternative name for watt-hour meter.

Integrator Cube, Sphere : Refers to a hollow cube, or sphere, the smooth interior surface of which is having a uniform flat, matt, while coating. If a lamp has been suspended within the cube or sphere, the illumination at any point of the surface (including a small observation window) would correspond to the rate of total light-flux emission.

Interconnected Star Connection : Alternative name for a rigzag connection.

Interconnector : Refers to a cable or line which is connecting two sources of energy or two distribution networks.

Interference : Refers to the undesired disturbance of signals in system which is caused by energy coming from an external source.

Interference Suppressor : Filter or attenuator fitted to apparatus which is capable of radiating interference signals, or to the terminals of the electric supply which might be a source of interference.

Intermodulation Distortion : Refers to a change in waveform, through the introduction of combination frequencies which are arising from two or more input frequency components, due to a non-linear response in the system.

International Ampere : Now absolete; see ampere.

International Ohm : Now obsolete; see ohm.

International Volt : Now obsolete; see volt.

Interphase Inductor : Alternative name for current sharing inductor.

Interpole : Refers to an auxiliary pole which is incorporated in a commutator motor. It is mounted on the magnet frame midway between the main poles and excited by a winding connected in series with the armature circuit. When the commutator moves relative to the brushes, reversal of current takes place in those coils connected to the commutator segments that are short-circuited by the brushes. This reversal gets delayed by the self inductance of the coils and the effect on the current-sharing, between the short-circuited segments, of the resistance current characteristic of the copper-graphite brush material. A delay in current reversal gives rise to bad commutation which is inducated by sparking at the trailing edge of the brushes.

Intertripping : Refers to the operation of all the circuit-breakers which are connected with a unit by using the functioning of a protective relay. It may relate to breakers situated in the same station or separated by many kilometres, as in the case of those associated with feeder transformers.

Intrinsic conduction : Refers to a conduction, in a semi-conductor by equal numbers of electrons and holes, *i.e.* by electron hole pair, as shown in Fig. 4. It is outweighted, in practice, by extrinsic conduction.

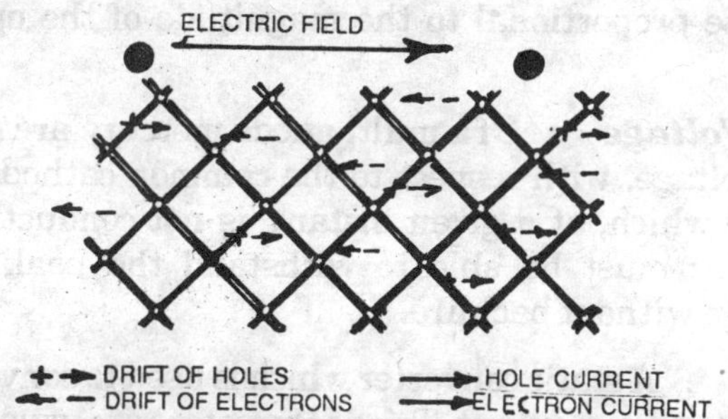

Fig. 4. Intrinsic conduction. Plane diagram of a diamond-type crystal lattice.

Intrinsic Impedance : Refers to the ratio of the electric to the magnetic field intensities in a plane electromagnetic wave in any medium or in space. For a given medium of conductivity ρ, permittivity ∈ and permeability μ, the intrinsic impedance for a sine-shaped travelling electro magnetic wave of angular frequency ω has been a follows:

$$Z_0 = \sqrt{[j\omega\mu/(\rho + j\omega\epsilon)]}$$

Intrinsic Strength : Refers to the maximum possible electic strength of a material. It has been the stress that would be withstood if breakdown from all causes (such as internal voids, contaminants and stress concentrations

due to electrode configuration) get removed. In practice this value has been never attained.

Intrinsically Safe : Term applied to apparatus of a circuit to signify that any sparking that may occur in it has been incapable of causing ignition of any specified flammable gases or vapours.

Inverse-speed Motor : Alternative name for a series-characteristic motor.

Inverse Square Law : A law of common occurrence in electrical engineering which applies to the force between point electric charges, the force between magnetic poles, intensity of illumination at a distance from light source, etc, the force or intensity in these instances varies inversely as the square of the distance.

Inverse Time Lag : Refers to a device which is able to delay the operation of equipment for a period which is inverse proportional to the magnitude of the operating force.

Inverse Voltage : Of a multi-anode mercury-arc rectifier, the voltage, with respect to the common cathode, of the anode which, at a given instant, is not conducting. The rectifier must be able to withstand the peak inverse voltage without backfire.

Inverter : A static converter which is used to convert a d.c. input to an a.c. output. Triode thyristors are usually used.

Ion : An atom or radial possessing a larger or smaller number of electrons than its normal content and consequently carrying a negative or positive charge. The conduction of charges in material bodies has been due to the motion of ions or electrons, or both.

Ionic Breakdown : Refers to a form of electric break down that takes place when the voltage stress gets raised to such a level that gas conduction occurs in small voids within the material.

Ionisation Chamber : Refers to a gas-filled enclosure which is used in particle physics. It consists of an enclosed vessel through which radiation has been directed, and in which there have been two electrodes. Commonly one is a central wire and the other a surrounding cylinder. This construction has been effectively a capacitor, and the presence of radiation reduces the effective insulation between the two electrodes so that more current will pass when a voltage has been applied. The current, which has been of the order of 1 pA, needs considerable d.c. amplification for its measurement.

Ionisation Current : The current due to the movement of ions and electrons that results from the action of an ionising agent in an electric field.

Ionisation Gauge : An instrument which is used for the measurement of high vacua. The electrons from a heated filament in a vacuum bulb get accelerated towards a collecting electrode, and the positive ions formed during their passage have been collected on a third electrode, the current recorded varying with the pressure inside the bulb.

Ionosphere : Refers to the upper part of the atmosphere where ionisation; mainly due to radiation from the sun, has been sufficient to have a marked effect on radio transmission.

I-pot : Alternative term for inductive potentiometer.

I.r. : Infra-red

PR Loss : Refers to the rate of energy loss as heat in a condcutor of resistance R carrying a Current I. The current I generally represents, for simple a.c. cases, the r.m.s. value and the resistance R must then be strictly constant or such value as would properly give the mean rate of energy loss.

Iron : A cheap and plentiful ferromagentic material (symbol: Fe; atomic weight 55.85; melting point 1535°C; boiling point 3000°C). In pure form it has been applicable to steady-flux applications such as machine frames; poles, yokes and rotors, electromagnet poles and yokes, relay cores, etc, where hysteresis and eddy-current losses have been immaterial. Its low resistivity has been a disadvantage in a.c. applications.

Iron Loss : Refers to the loss in an iron material due to varying magnetisation.

Irwin Oscillograph : A mechanical oscillograph which uses a thermal element. It has been otherwise similar in operations to the Duddell oscillograph.

Isolator, Isolating Switch : A switch which is used for isolating a circuit or disconnecting it from the supply, which could be operated only when no current has been passing through it.

Iterative Impedance : Of a two-port network, that impedance which, when connected across one pair of terminals, produces a like impedance across the other pair.

J : Refers to the mathematical symbol which is used to signify the imaginary component of a complex quantity.

Jar : A unit of capacitance which is once used in the Royal Navy. It has been euqal to 1/900 μF. The name has been derived from the Leyden jar capacitor.

Jitter : A temporal irregularity in the repetition of a well-defined pulse.

Johnson Noise : Obsolete term for thermal agitation noise.

Joint Box : See junction box.

Jointing Chamber : A chamber at which cables could be jointed.

Joubert Disc : Refers to a device applied, in a point-by-point measurment of the amplitude of a voltage,.at a series of selected instants in a repeated cycle. It includes a circular plate of insulating material, carrying a narrow metallic insert at one point on the rim. The disc has been rotated synchronously. The insert then forms an intermittent contact from which a sample voltage could be picked off at the same instant in successive cycles of a given wave.

Joule : It is the SI unit of energy irrespective of whether mechanical, electric or thermal (symbol : J).

Joule Effect : Refers to the generation of heat by an electric current in a material at a rate which is proportional to the square of the current density and the resistivity of the material. The rate of heating in a resistor of resistance R carrying a current J has been PR.

Jumper : In general, this term refers to flexible length of conductor which is used to connect two fixed terminations; it facilitates a ready alternation of the interconnection when required. In an overhead line, it refers to a non-tensional conductor connected across two anchor clamps to maintain electric continuity. In electric traction, a jumper cable has been an insulated cable which is connected across the gaps in a conductor rail for the same purpose.

Junction Box : Refers to a closed box which is usually underground,to which have been brought the ends of feeders and distributing mains for connection and protection.

Junction Diode : Refers to a semiconductor diode in which the 'anode' and 'cathode' have been in large-area intimate contact.

Junction Law : Refers to a primary law of electric networks which expresses the continuity of current.

Junction Transistor : Transistor having either a zone of n-type semiconductor sandwiched between to p-type zones, or a zone of p-type semiconductor sandwiched between two n-type zones.

Kalium Cell : Refers to a dry primary cell having a mercuric oxide/potassium hydroxide/zinc system. It will hold its voltage for long periods and at high temperatures.

Kaplan Turbine : Refers to a type of propeller turbine in which the blades are having an adjustable pitch enabling them to be accommodated to a wise range of operating conditions. Kaplan turbines find use in large sizes for heads of water as low as 6m.

Kapp Vibrator : Refers to a set of three d.c. type commutator armatures having permanent-magnet fields, each connected into one rotor phase circuit of a slip-ring polyphase induction motor for raising the power factor.

Keeper : A piece of ferromagnetic material which is kept across the extremities of a permanent magnet to complete the magnetic circuit.

Kelvin : The SI unit of thermodynamic temperature (symbol: K).

Kelvin Balance : A electrodynamic measuring instrument in which the electromagnetic forces gets balanced against the force of gravity using a travelling weight.

Kelvin Double Brdige : Refes to an adaptation of the Wheatstone bridge which finds use for the measurement, of low resistances and four-terminal shunts. From Fig. 1, the currents in S and R have been equal at balance, so that

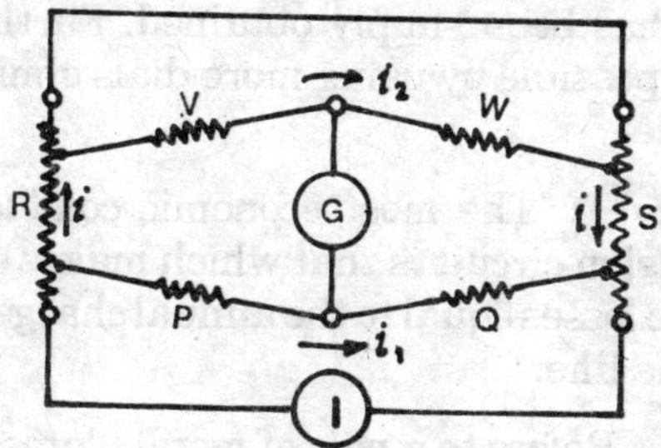

Fig. 1. Kelvin Double Bridge

$$Si=Qi_1-Wi_2;\ Ri=Pi_1-Vi_2$$

If

$$W/Q=V/P$$

then

$$S/R=Q/P=W/V$$

The bridge gets balanced with Q=W both fixed, and P and V varied together.

Kelvin Effect : See skin effect.

Kelvin - Varley Slide : Refers to a form of electric vernier, which consists of a set of coils and a stud swtich taking off a small fixed fraction of the overall voltage, which then feeds another set of coils having a single tapping, in the two-dial form (see Fig. 2). The top dial is having 101 coils each of 100 Ω and the lower is having 100 coils of 20 Ω totalling 2000 Ω. The lower coils bridge two coils in the first dial so that the total resistance has been effectively equal to 10000 Ω. In this way sub-division to

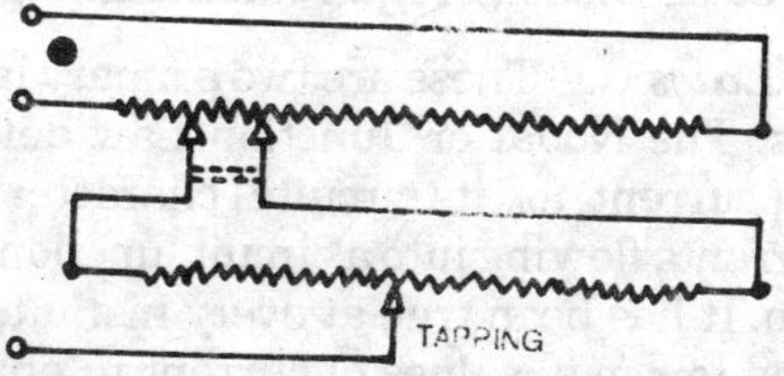

Fig. 2. Two-dial Kelvin-Varley slide.

1/10000 of the main dial resistance (and potential difference) has been simply obtained. Further subdivision becomes possible by using more dials similar to the first, in cascade.

Kelvin's Law : The most economic conductor area for a transmission circuit is that which makes the annual cost of the line losses equal to the annual charges on the capital cost of the line.

Kerr Cell : Refers to a pair of metal plates, which forms a capacitor and gets sealed into a glass tank filled with nitrobenzine. If a beam of plant-polarised light is allowed to pass through the cell between the capacitor plates, with the cell flanked by a pair of crossed Nicol prisms, the field of view has been dark. When a potential difference is applied to the capacitor, establishing an electric field across the path of the light, the plan of polarisation gets rotated and the field of view brightens.

Keyholder : A lampholder which is incorporating a switch. It is also termed as a switch-lampholder.

Kilowatt-hour : A practical measure of energy. It may be defined as the energy expended in one hour when the power is 1000 watts. It is equal to 3.6×10^6 joules in SI units.

Kilowatt hour Meter : = See integrating meter.

Kiosk Switchgear : Refers to the outdoor counterpart of cubicle switchgear. It finds use mainly for suburban distribution systems. It consists of a sheet-steel weatherproof kiosk which houses a step down transformer with oil circuit-breaker for switching and protection, and a number of fuse-protected outgoing circuits for lower-voltage distribution, The kiosks gets replaced or transferred as the load requirements change.

Kirchhoff's Laws : These are two primary laws of electric networks. The Nodal or Junction Law defines the continuity of current, *i.e.* its circuital characteristic; the sum of the currents flowing into a circuit junction (or node) has been zero. It has been true at every instant, and can also be applied to r.m.s. values of current in phasor form. Its statement has been the follows :

$$\Sigma i = 0$$

with due regard to sign; *i.e.* if currents flowing towards the node are considered as positive, then those flowing away from it are negative. The Mesh Law defines the principle of energy conservation: the total source e.m.f. in a closed circuit (or mesh in a network) would be equal to the sum of the voltage drops round the mesh, or

$$\Sigma e = \Sigma iZ$$

In a single circuit with current i everywhere the same, ei denotes the electric output of the sources, and i^2Z the total rate of energy absorption by the rest of the circuit, and these must be equal.

Klydonograph : An apparatus on which a Lichtenberg figure gets recorded. It is made with multiple electrodes and with provision for both polarities. It records for voltage of 3—15 kV.

Klystron : Refers to a velocity-modulated thermionic tube, using the finite transit time of the electron in producing very-high-frequency oscillatory power. It consists of an electron gun (emitter cathode and focusing electrode), two tuned cavity resonators and a collector electrode.

Kraemer System : A system which is used for regulating the speed of a large induction motor by employing as an auxiliary machine a synchronous converter,to change the slip power from slip frequency to zero (*i.e.* direct current).

Korndorfer Starting : Refers to a method of starting motors which avoids the interruption of current that takes place with the simple auto-transformer starter. This is attained as shown in Fig. 3. On the first step, (a), the motor gets accelerated at a reduced voltage determined by the transformer tapping. On the second step, (b), the star point of the transformer gets opened so that the motor continues to run with part of the transformer winding in circuit. Next, this part gets short circuited by the 'run' contactor or switch, and finally the 'start' contactor or switch gets opened, (c).

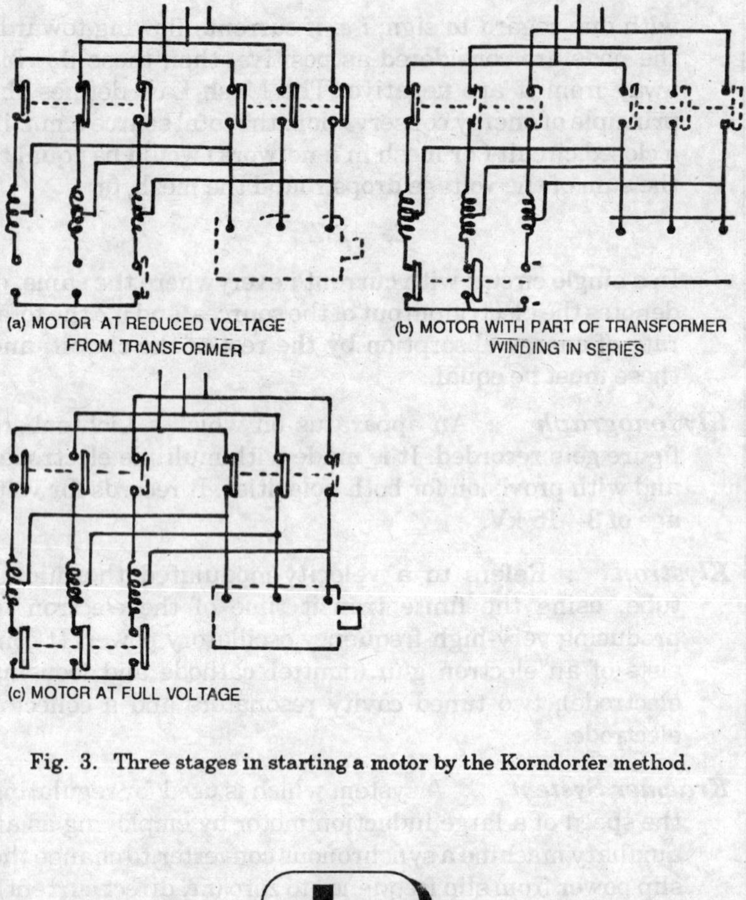

(a) MOTOR AT REDUCED VOLTAGE FROM TRANSFORMER

(b) MOTOR WITH PART OF TRANSFORMER WINDING IN SERIES

(c) MOTOR AT FULL VOLTAGE

Fig. 3. Three stages in starting a motor by the Korndorfer method.

Lag : Refers to the fraction of a period which is generally expressed as a corresponding fraction of the angle 2π, by which a given sinusoidal function (*e.g.* an alternating current) gets displaced behind a reference function (*e.g.* an alternating voltage) of the same frequency.

Lagging Load : Refers to a reactive load in which the current at the terminals gets lagged in phase behind the voltage at the same point. It is sometimes known as an inductive load.

Lagging Phase : In the two wattmeter method of three-phase power measurement, the conductor in which the current at unity power factor lags behind the voltage associated with it in the wattmeter. The term is also applicable to a three-phase circuit to indicate the phase whose voltage is lagging behind that of one of the other phases by 120°.

Lambert : Refers to the non-rationalised metric unit of luminance (symbol: L). The SI unit of luminance has been the candela per square metre (symbol: cd/m^2); thus $1L=10^4/\pi$ cd/m^2.

Lamination : Of a machine or transformer, this term refers to each of the thin, steel sheets which form part of the core. The Sheet are also known as core plates, punchings, or stampings.

Lamp : Refers to a means of converting electric energy into light. There have been three practical ways of doing this: (a) by using the heating effect of an electric current which raises a fine filament of wire to a state of inacandescence so producing light (see filament lamp), (b) by an electric current through a gas which creates characteristic colours of light (see discharge lamp), and (c) by a two-stage process whereby ultraviolet radiations have been first generated and these in turn get changed into light rays by the process of fluorescence (see fluorescent lamp).

Landing Switch : Refers to a single pole change over switch having no 'off' position, for use in a circuit that needs control from two or more positions. It is also known as a two-way switch.

Lap Welding : Refers to a form of resistance welding in which the parts being joined get overlapped. Varieties of lap welding have been as spot welding, projection welding, seam welding and roller-spot welding.

Lap Winding : Refers to a spread form of winding in fan electric machine in which successively connected coils, of approximately full pitch, are over lapping each other.

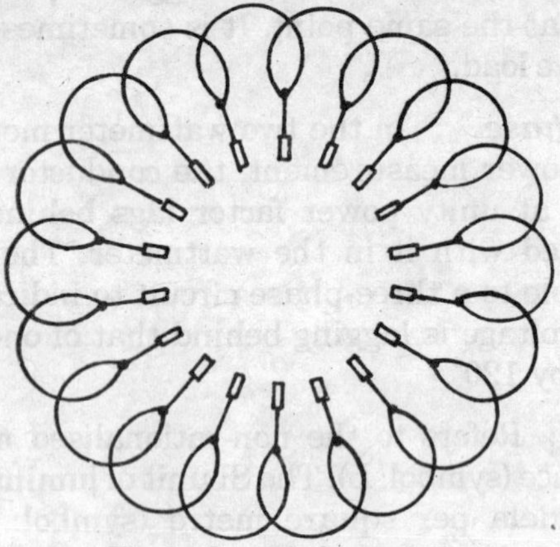

Fig. 1. Lap-wind armature.

Larsen Potentiometer : An arrangement which is used for the measurement of alternating potential differences. In Fig. 2, where E denotes the unknown voltage, the primary winding of a variable mutual inductor and a resistor provided with tappins get connected in sereis to a source of supply E, E_x, and E have been generally derived from the same source since they must, of course, be of the same frequency. The current I may get regulated by a rheostat R_1. The induced e.m.f. in the secondary of ther mutual inductor in series having a tapped fraction of the potential difference on the resistor R now gets balanced against the unknow E_x. The detector has been

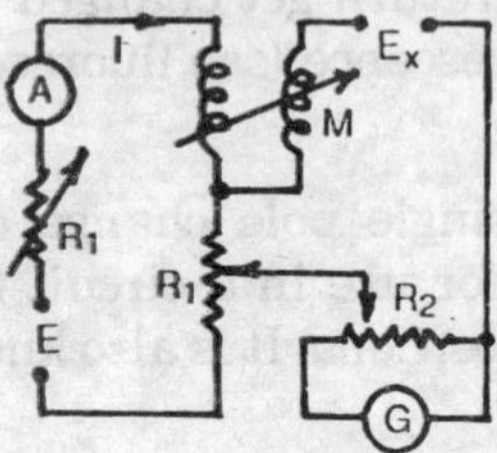

Fig. 2. Larsen Potentiometer.

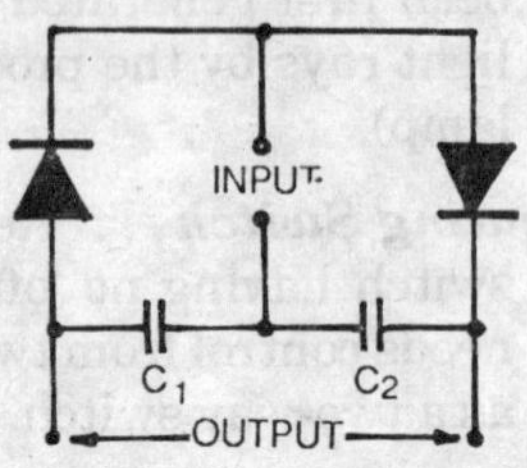

Fig. 3. Latour doubler

conveniently a vibration galvanometer G, whose sensitivity could be varied using R_2. If M and R denote the readings of mutual inductance and resistance at balance $Ex = I\sqrt{[R^2+(\omega M)^2]}$, because the components IR and $I\omega M$ have been in quadrature. The phase angle ϕ between E_x and I has been $\phi = \arctan(\omega M/R)$.

Laser : Refers to an electronic device which has been named ith the acronym of Light Amplification by Stimulated Emission of Radiation. It emits light as a coherent wave in a monochromatic beam of high intensity. Lasers could be obtained from solid, gaseous or liquid materials. Certain types of laser emit pulsed light; the light emission from other types has been continuous and may get modulated in intensity after emission. Applications of appropriate types of lasers have been in a variety of fields like high-precision as well as micro-machining and welding of metals, in surgery, in chemistry, in measurement of distances and in telecommunication.

Latching Relay : A relay which needs the electric or mechanical actuation of a latch to release the contact units after its operation.

Latour Doubler : Refers to a basic form of voltage doubler. The a.c. input has been applied to the centre of the circuit in Fig. 3. The potential differences across C_1 and C_2 each equal the peak input voltage. Hence the output voltage across the two capacitors has been double the peak input voltage.

Lattice Diagram : Refers to a graphical construction which is used for the rapid assessment of surge propagation and reflection on a transmission network.

Lattice Network : Refers to a form of passive two-port network.

Lattice Tower : Refers to a structure of metal bars to support, at a suitable height from the ground, the phase conductors of an overhead line. It is popularly termed as a pylon.

Lattice Winding : Distributed winding in which the overhang has been arranged to give diamond-shaped coils.

Lay : refers to the axial length of one turn of the helix which is formed by a cable core or the strand of a conductor. The term is sometimes used to mean lay ratio.

Lay Ratio : Refers to the ratio of the lay of a cable or conductor to the mean diameter of the helix.

L.D.F. : Load factor.

Load : The term used for the fraction of a period by which a given sinusoidal function (*e.g.* an alternating current) gets displaced in advance of a reference function (*e.g.* an alternating voltage) of the same freqency. It is generally expressed as the fraction of the angle 2π.

Lead-acid Cell : Refers to a storage cell or accumulator in which the positive plates have been of lead peroxide (PbO_2), the negative plates pure lead in spongy form, and the electrolyte dilute sulphuric acid (H_2SO_4). During discharge the lead peroxide partially gets reduced the spongy lead gets oxidised and both products react with the sulphuric acid to produce water and lead sulphate. During recharge the process gets reversed:

$$PbO_2+2H_2SO_4+Pb \longleftrightarrow PbSO_4 \longleftrightarrow +2H_2O+PbSO_4$$

Formed plates and pasted plates are used. The nominal voltage of a lead-acid cell has been 2 V.

Lead-in Insulator : Alternative name for bushing.

Leader Stroke : Of lightning this term means the discharge which initiates the track of the lightning channel.

Leading Load : Refers to a reactive load in which the current at the terminals leads in phase the voltage at the same point. It is sometimes known as a capacitive load.

Leading Phase : In the two wattmeter method of three-phase power measurement, this term refers to the conductor in which the current at unity power factor leads the voltage associated with it in the wattmeter. The term is also applicable to a three-phase circuit to indicate the phase whose voltage is leading that of one of the other phases by 120°.

Leakage Current : Refers to a small fault current.

Leakage Flux : Refers to the part of the flux which is produced by a magnetic field that traverses non-useful paths. Leakage flux affects the field excitation needed by d.c. excited windings, and the leakage reactance of a.c. windings.

Leakance : Refers to the reciprocal of insulation resistance. It is also termed as leakage conductance.

Leblanc Advancer : Refers to a phase advancer which consists of an armature having a commutator and three brushes per double pole-pitch spaced 120 electrical degrees. The brushes get connected to the slip rings of the main induction motor. The machine produces an electromotive force which is approximately proportional to the rotor current of the induction motor, and leading or lagging on it by 90° depending on the speed at which the armature has been driven. No stator winding is required but a stator core may be provided to reduce the reductance. Commutation has been difficult, and as a consequence the advancer has been limited in size.

Leblane Connection : The term used for a method of interconnecting three single-phase transformers or windings which is used for the conversion of three-phase voltages to two-phase, or two-phase voltages to three-phase.

Leclanche Cell : A primary cell, originally wet, which is having zinc and carbon electrodes in a solution of ammonium chloride, with a manganese depolariser. A dry form of construction is more common in which the electrolyte has been a thick paste of jelly or starch (see Fig.5). In its cylindrical form, it finds main use for torches.

Leg : Of a transformer, this term refers to that part of the core which is sorrounded by the windings It is also known as a limb.

Lenz's Law : If an electromotive force gets induced in a circuit by a changing magnet flux, the direction of the induced voltage would be such that any current produced by it opposes the change of linkage.

Leyden Jar : A capacitor having a jar, generally of glass, having its inner and outer surfaces coated with a conductive material.

Lichtenberg Figure : A figure which is produced on a photographic plate separating two electrodes when a surge voltage has been applied. The normal figure obtained (if the peak voltage is high enough) consists of a set of radial streamers, with a well-defined radius which has been a measure of the peak voltage.

Life Test : A test on a component or device which is used to establish its probable life under specified conditions of operation.

Lift Machine : Refers to the part of an electric lift system which includes motors, any reduction gear, brakes and winding drum or sheave. Modern electric lifts include mainly 'geared' or 'gearless' traction machines. Winding drums are confined to lifts having a low rise.

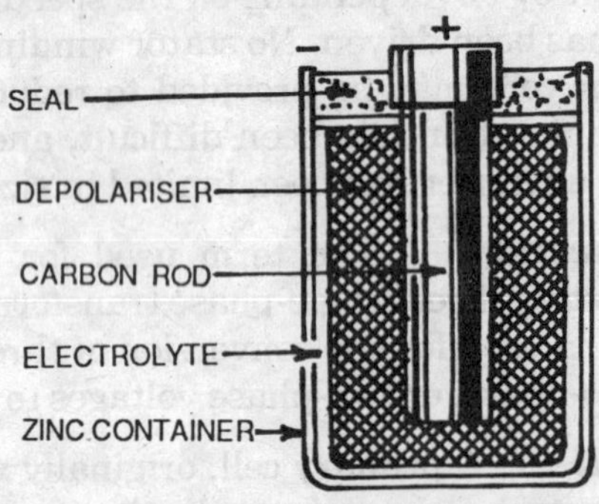

Fig. 5. Dry Leclanche Cell.

Lifting Magnet : An electromagnet which is used in industry for handling iron and steel. For general lifting duties, the circular magnet has been most common, while, plate handling and the moving of long pipes has been done with a rectangular magnet; a multi-polar type is used for lifting coiled sheet metal.

Light-emitting Diode : It is a semiconductor diode which is usually of gallium arsenide, whose p-n junction would emit light at a fixed frequency when it carries a current of a sufficient magnitude.

Lightning : Refers to an electric discharge which occurs between two clouds charged to different potentials or between a charged cloud and the Earth's surface.

Lightning Arrester : A surge diverter which is used to protect apparatus from surges arising from lightning.

Lightning Conductor : Refers to a continuous system of metallic conductors which is arranged to provide easy passage to earth from the highest point of a building for a lightning discharge. It consists of one air termination and one down conductor.

Limb : Of a transformer, this term refers to that part of the core which is surrounded by the windings. It is also known as leg.

Limit Switch : A switch which is fitted to movine equipment like a lift or travelling crane to cut off the power supply if the equipment tries to move beyond a particular point. It finds use either as a safeguard against errors of judgement by an operator or as a necessary feature of automatic operation. If it is used to break the main supply it is termed as a series limit switch; but when it operates in a control circuit cutting off the supply by the action of a contactor or circuit-breaker, or initiating some protective altered action by energising a trip coil or relay circuit, it is termed as a shunt limit switch.

Lindemann Electrometer : An electrometer which is having a very small and light moving system providing a very short periodic time. The reading system comprises of a microscope. The quadrants of a basic electrometer have been replaced by pairs of slotted flat plates, and instead of a vane there has been a metallised quartz fibre carried on a similar fibre under tension to provide connection and constraint.

Line-drop Compensator : A device which incorporates resistance and reactance. It is used to maintain feeder voltage constant at a point remote from the regulated transformer. It allows to pass a current proportional to the load current and produces at the control point a voltage which is equal to, and in phase with, the drop along the feeder.

Line Inductor : Refers to an inductor which is connected in series with electric equipment to absorb (or reflect) the effects of high-frequency surges. The alternative term line choke gets deprecated.

Line Voltage :

(1) Refers to the voltage between the two lines of a single-phase system.

(2) Also, refers to the voltage between any two lines of a symmetrical three-phase system.

(3) Also, refers to the voltage between any two consecutive lines of a symmetrical six-phase system.

Linear Acceleratar : Refers to an indirect particle accelerator in which the particles travel in a straight line and arrive at gaps in the structure at the right phase of an r.f. excitation, or move in steps with a travelling electromagnetic wave. Acceleration of protons is achieved by passing them through a resonant cavity having a series of draft tubes, which is spaced and dimensioned in such a way that the electric field between gaps nearly remains constant. Particles get accelerated between gaps and move between centres of successive gaps in one complete cycle of oscillation. An important machine for the acceleration of electrons has been the travelling wave accelerator which is using megawatt pulses of h.f. power at 3000 MHz. The power gets propagated along a 'corrugated waveguide', a section of circular waveguide containing a series of irises. The travelling wave set up is having an axial component of electric field which is used to accelerate the electrons.

Linear Circuit, Netuork : Refers to a circuit or network which comprises only such circuit elements for which the voltage/current charateristics have been linear relationships.

Linear Motor : A motor which has been designed to produce linear, rather than rotary, motion. It may be viewed as a conventional annular three-phase stator cut across at one point and laid out flat, the 'rotor' (now moving in a straight line) being an equally conventional magnetic inductor. Induction machines have been most common, but as many types of linear motor have been possible as there have been rotary machines.

Linkage : See flux-linkage.

Liquid-quenched Fuse : A fuse which uses a liquid to quench the arc. It is usually a cartridge fuse, and could be referred to as simply a liquid fuse.

Liquid Starter, Controller. : Refers to a rheostatic starter or controller in which the resistor has been liquid.

Lissajous Figure : Formarly a closed graphical figures which are formed by the reflection of light rays from vibrating tuning forks; now generally refers to that formed on the screen of a cathode-ray tube by the combined action of two cyclic voltages which are applied to pairs of deflecting plates arranged at right-angles.

Litzendraht : Stranded cable which is used for high-frequency currents. In order to reduce skin effects, individual strands have been insulated and so woven that each occupies all possible geometrical positions in the cable section within a comparatively short overall lenght. It is also termed as litz wire.

Live : Term used for an object when either a difference of potential exists between it and earth, or it gets connected to the 'neutral' of a supply system in which that conductor does not get permanently and solidly earthed.

Live-line Indicator : It is an instrument which is used for indicating whether an overhead transmission line is live, disconnected (but carrying a charge) or dead. The live-line tester usually consists of a relatively low-range electrostatic voltmeter or rectifier-type voltmeter having a high-insulation capacitor in series, the combination being mounted near the end of a long insulated pole which can get raised to the line from the ground, making a safe reading to be made by a user on the ground.

Lloyd-fisher Square : Refers to an assembly of strips of ferromagnetic material which could be used for the measurement of core loss. The strips get arranged on edge in the form of a hollow square and interlinked with corner pieces.

Load : Refers to the power output of a generator, motor, transformer, etc, or the power carried by a circuit.

Load Angle : The term used for the electrical angle between the stator and rotor magnetomotive forces in an electric machine. It has been significant in all rotating machines, but its effects hve been most usually observable, and operationally important in synchronous machines. It is sometimes termed as the power angle or torque angle.

Load Curve : Refers to a graph in which power supplied or consumed gets plotted as a function of time.

Load Discriminator : A device which is used for diverting the series winding of a crane motor to get the optimum speed/torque ratio and the maximum acceleration.

Load Factor : May be defined as the ratio of the number of energy unit which are supplied during a given period to the number that would have been supplied had the maximum demand been maintained throughout the period.

Locked-coil Conductor : Refers to a stranded conductor having its outer wires so shaped that they get restrained from any radial movement.

Locus Diagram : A curve which is plotted on rectangular or polar co-ordinates, to show the variation of a system function (*e.g.* admittance, load, transfer function) when some parameter other than time gets changed in steps. Each point on the curve denotes a sinusoidal steady state for a given set of parameters, transient condition being ignored.

Logarithmic Decrement : Refers to the natural logarithm of the decrement of an oscillating system.

Logic Element : Refers to a device where the values of two or more input signals give the value of one or more output signals.

Lohys : A dynamo steel which is used for the cores of general-purpose industrial motors. It is having a silicon content of 0.2%.

Loop Test : It is a localising a fault in a cable. The faulty conductor get looped to a sound conductor. Variations include the Murray loop test, Werren overlap test, open and closed test, Fisher loop test and Hilborn loop test.

Loop Winding : Refers to a method of winding a lap-connected armature.

Looping-in : Refers to a method of avoiding the use of a T-joint in wiring installation. The conductor has been taken to and from the point to be supplied.

Loss Losses : Of a device, equipment of plant, this term refers to the diffrerence between energy input and useful energy output.

Loss Angle : Refers to the angle by which the angle between the leading current and the voltage in a piece of insulating material is different from 90°. It has been equal to zero for a perfect insulant.

Loss Factor : Of a particular type of loss like the I^2R loss, this term refers to ratio of the actual number of units consumed in this loss over a given period, to the number of units that would have got consumed if the maximum loss had been maintained continuously over the whole period.

Loss Tangent : Refers to the tangent of the loss angle in insulating material. It has been approximately equal to the power factor.

Low-frequency Induction Furnace : Refers to an induction furnace that has been particularly adapted to non-ferrous metals, especially brass and, more recently,

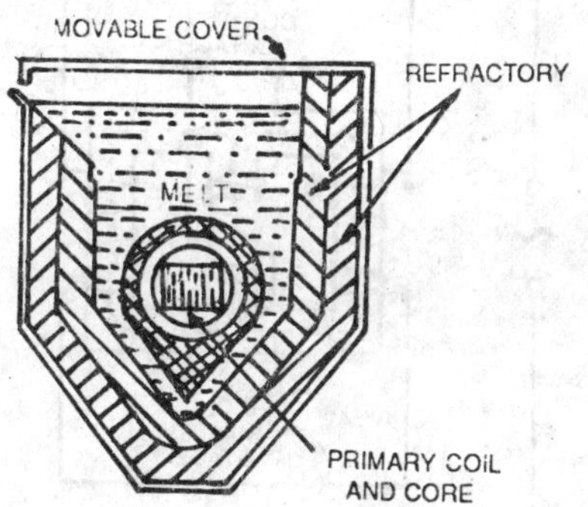

Fig. 6. Low-frequency Induction Furnace.

aluminium alloys. It comprises of a single or double-yoke core with a primary winding in its centre to which current has been supplied at 150-60 Hz. The hearth is having a narrow channel in the ower part where melting occurs.

Low Voltage A term used to imply a voltage not exceeding 250 V.

Lumped Circuit, Network Refers to a circuit or network that comprises no distributed circuit element.

Magnetic Amplifier : Refers to a device where a transductor has been used to achieve amplification.

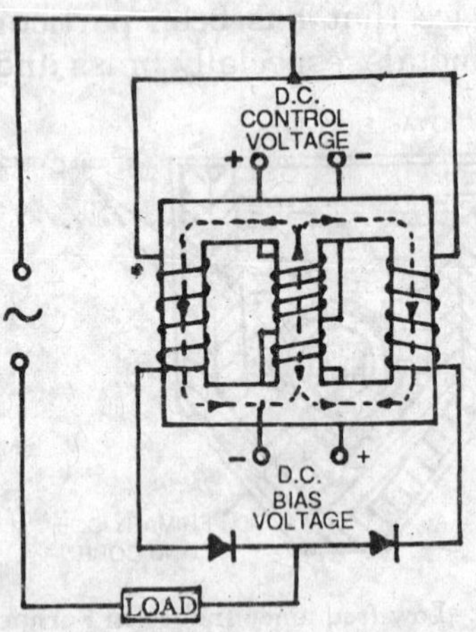

Fig. 1. Basic circuit for a full-wave magnetic amplifier. The iron-cored coil forming the basic amplifying element, is termed as a transductor.

Magnetic Blowout : The earliest type of arc-control device. By using a coil carrying the current to get interrupted, it sets up a magnetic field across the arc. It gets accentuated the natural tendency for the arc to lengthen, by using the electromagnetic forces resulting from the interaction between the field and the arc current.

Magnetic Brake : In its simplest form, it is an a.c. or d.c. solenoid which has been ranged to operate, through a system of levers, a pair of brake shows provided with friction linings. When the solenoid get deenergised, the brake wheel gets held stationary by the brake shoes under the action of a counterweight or, more usually, a spring. But when the solenoid get energised, the brake shoes get lifted against the spring pressure and the wheel is free to more.

Magnetic Circuit : Refers to the closed path round which magnetic flux passes. Megnetic flux could not get confined as can an electric current: there has been no magnetic insulator.

A circuit law is relating to total magnetic-circuit flux ϕ the magnetomotive force F producing it through a quantity called the reluctance S, in the form $\phi = F/S$. This expression has been of rather limited application, because the reluctance has been a function of the absolute permeability, which in useful ferromagnetics has been variable.

Magnetic Clutch : Clutch usually having two parts: a field member having an electromagnet and the necessary slip rings through which the current gets supplied to the coil, and an armature member. It is usual for the field member to get mounted on the driving shaft and the armature on the driven shaft. The two halves of the clutch have been held apart by using a spring plate when the coil winding is not energised.

Magnetic Constant : The quantity μ_0 which in the SI is having the value $\mu_0=0.4\pi$ microhenery per metre. It is also termed as absolute permeability of vacuum and gets related to the electric constant ε_0 and the speed c_0 of light by the expression

$$\varepsilon_0\mu_0 c^2_0=1$$

The magnetic constant is having the function for relating the electric and magnetic units of a system of units of measurement to the mechanical units of that system.

Magnetic Difference of Potential : Refers to a difference in the magnetic states existing at two points which generates a magnetic field between the two points. It would bc equal to the line-integral of the magnetic field strength between the two points, except in the presence of electric currents.

Magnetic Field : Ther term used for the state of the space in the vicinity of an electric current or a permanent magnet throughout which the forces produced by the current or magnet have been discernible.

Megnetic Field Strength : At a point in a magnetic field this term refers to the quotient of the magnetic flux density by the absolute permeability of the material. Its SI unit has been the ampere per metre (symbol: A/m).

Magnetic Flaw Detection : Refers to a method of detecting flaws in metal castings; it is especially effective for checking for surfce cracks on ferromagnetic material. It's working depends for its application on the leakage of magnetic flux from the surface of the specimen in the vicinity of a crack-like discontinuity. Near the crack, lines of flux would pass through the air and any small magnetic particles near the edge of the crack will, therefore, get collected in this region and reveal the presence of a crack.

Magnetically Hard Material : A magnetic material having a high value of coercivity.

Magnetical Soft Material : A magnetic material having a low value of coercivity.

Magnetic Flux : The term used for the concept of the magnetic properties of a magnet as appearing to 'flow' along definite paths called lines of magnetic force, the total number of such lines of force issuing from a particular magnet pole being the magnetic flux. Its SI unit has been the weber. If the flux linking one turn in a circuit alters by one weber in one second, a voltage of one volt will get induced in that turn. The c.g.s. unit has been the maxwell; one weber=10^3 maxwells.

Magnetic Hysteresis : Refers to a phenomenon which is observable in the magnetisation of some materials in which their magnetic state at an instant gets related to their previous state.

Magnetic Leakage : Refers to the part of a magnetic flux that would follow a path in which it has been ineffective for the purpose desired.

Magnetic Link : A device that reveal the magnitude of a lightning surge through a conductor. It is generally in the form of a container, holding highly remanent steel wires or laminae, which gets mounted adjacent to the conductor, and indicates by a change in its magnetic state the passage of the lightning current. It is also termed as a surge-current indicator.

Magnetic Moment : Of a magnet situated in a uniform field in a vacuum this term refers to the ratio of the torque on it when in the position of maxiumum torque to the magnetic flux density of the field.

Magnetic Neutralisation : The term used for the process of bringing a magnetised body or substance to the neutral magnetic state.

Magnetic Overload Relay : Refers to an overload relay which is using the magnetic effect of the current for operation. When the current flowing in the coil gets reacted a predetermined value, a plunger housed in the dashpot gets attracted into the coil. When near the end of its travel, it opens the normally closed trip contact. In order to get a time delay, the dashpot would be filled with a suitable fluid, usually a mineral oil.

Magnetic Separator : A device which is used for removing ferro-magnetic materials from a mixture. It consists of an electro-magnetic and may be used when material has to be handled in bulk.

Magnetic Sheet Steel : It constitutes basic constructional material of all electric motors, generators and transformers. It is iron alloyed with 0.4 – 4.2% silicon, rolled into sheets or strips whose thickness usually lies between 0.35 and 0.63 mm. Some times a small amount of aluminium is also added, but in general, other additional elements get reduced to very small proportions. In particular, carbon has been a serious cause of increased magnetic losses; it is usually reduced to 0.02% in hot-rolled, and below 0.005% in grain-oriented, sheets.

Magnetic Shunt : Refers to a component of magnetic material which is used to divert magnetic flux to a parallel section of the magnetic circuit.

Magnetic Space Constant : Deprecated term for magnetic constant.

Magnetic Storm : The term used for a substantial temporary disturbance of the magnetic field of the Earth which is caused by an abnormal influx of charged particles from the sun. It is able to disrupt some radio communications and induces fluctuating currents in power and telephone lines.

Magnetic Tape : Refers to a tape or ribbon which is coated with a thin film of magnetic material used, for example, to record voice or video signals, or digital data in data processing.

Magnetisation : Refers to the process of producing in a magnetic body the characteristic properties of a magnet.

Magnetisation Characteristic : Refers to the relation between the magnetic flux density and field strength for a prepared specimen of magnetic material; or the relation between the induced voltage on no load to the field (magnetising) ampere-turn excitation, for a given speed. The latter is often termed as the open-circuit characteristic.

Magnetising Coil : See field coil.

Magnetising Force : Deprecated term for magnetic field strength.

Magneto : Refers to a magneto-electric generator, in which the magnetic flux has been provided by permanent magnets. It has been capable of producting high-voltage pulses and has been sometimes used in the ignition circuit of internal-combustion engines.

Magneto-Electric Generator : See magneto.

Magnetohydrodynamic : Refers to the behaviour of electrically-conducting gases or liquids in the presence of electric and magnetic fields.

Magnetohydrodynamic Generator : A device which is used for converting thermal energy into electric, by braking a stream of hot ionised gas. In the simplest type of m.h.d. generator, the gas stream is allowed to pass through the poles of a magnet at right-angles to the magnetic field. With suitable pick-up arrangements, the electric field induced in the stream may find use to drive a current through an external load. The device is also called a plasmahydrodynamic generator.

Magnetometer : An instrument which is used for the measurement of a weak magnetic field (originally the Earth's field).

Magnetomotive Force : Refers to the cause of the existence of a magnetic flux in a magnetic circuit. Quantitatively, it refers to the scaler integral of the magnetic field strength along a closed path.

Magneton : The term used for the magnetic moment of a spinning electron. A rotating charge has been equivalent to a circular current and gives rise to a magnetic field.

Magnetoplasmadynamic Generator : See magnetohydrodynamic generator.

Magnetostatic Lens : See electron lens.

Magnetostriction : The term used for the phenomenon whereby, when a magnetic material gets magnetised, its dimensions gets changed. Conversely, if the material gets strained, changes take place in its magnetic properties. Magnetostrictive effects include the Villari effect and the Wiedemann effect.

Magnetostrictive Vibrator : A vibrator which uses the phenomenon of magnetostriction. A low-frequency magnetostrictive vibrator is depicted in Fig. 2.

Magnicon : A form of rotating amplifier having cross-field excitation.

Magslip : Refers to a small precision-built electric synchro which is applied to position-indication and comparable functions in automatic control systems.

Magnetron : Refers to a value which is used for generating power at centimetric wavelengths. It uses the electron transit time between internal elements which take the form of reasonant cavities.

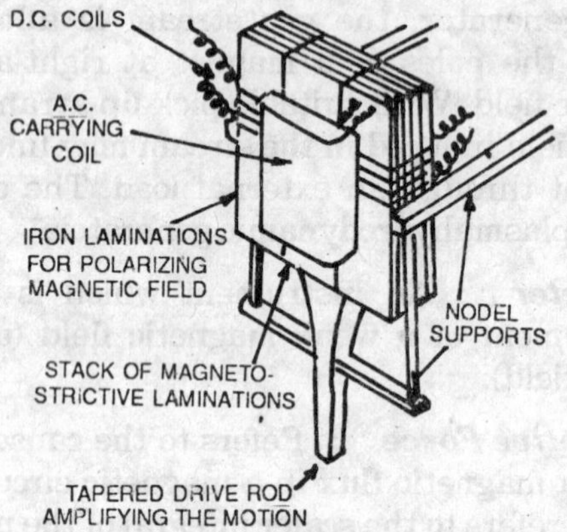

Fig. 2. Megnetostrictive vibrator (W. Bryan Savage Ltd.)

Main : Conductor or conductors which are arranged for the transmission and distribution of electric energy.

Making-current : Of a switch, circuit-breaker, etc., this term refers to the total maximum current peak which takes place immediately after the circuit gets closed.

Manganin : Refers to a copper-manganese-nickel alloy having high resistivity and low temperature coefficient of resistance. It frequently finds use in the manufacture of instrument resistors.

Manipulator : Equipment which is able to simulate the actions of a human arm and hand, for use in situations where operations should be carried out remotely, for example on toxic or radio-active material.

Maser : A device which is based upon the fact that an elemental atomic magnet in a magnetic field may be having one of two energy states, *i.e.*, parallel or anti-parallel to the field. If electro-magnetic radiation of suitable frequency gets applied to the system, transitions of atomic magnets from one energy state to the other can occur, energy being absorbed if the transition occurs from the lower energy state to the higher, or radiated if the transition occurs from the higher to the lower. Employing this principle, practical amplifiers and oscillators have been produced. The name has been derived from the term

'Microwave Amplification by Stimulated Emission of Radiation'.

Mass-impregnated Cable : Refers to a paper-insulated cable in which the paper tapes have been applied unimpregnated, the complete cable being subsequently dried and impregnated with compound as a whole.

Mass-impregnated Gas-pressure Cable : A cable in which the normal mass-impregnated insulant is used but the lead or aluminium sheath is applied with a small clearance. This space between the screened insulant and the sheath has been charged with nitrogen gas at high pressure. In this design the gas is allowed to pass into and out of solution with the compound with variations of temperature.

Mass Resistivity : May be defined as the product of the volume resistivity and the density of a material at a given temperature.

Mass Spectograph : Refers to an electronic device which is so designed that a beam of a mixture of ions passes through a uniform magnetic field of constant flux density, which is able to separate the ions according to their different masses. It has been used for the chemical analysis of gaseous mixtures.

Mass Susceptibility : Refers to the quotient of susceptibility and magnetic flux density.

Mass-type Plate : Of an accumulator cell, a plate having large blocks of active material held in a frame.

Master Controller : Refers to a multi-way switch for controlling the operation of a set of contactors. It is sometimes termed as a pilot controller.

Master Frequency Meter : Refers to an instrument which sums the total number of cycles of an a.c. supply in a given time, thus making a comparison to be made with the required number. It is sometimes termed as an integrating frequency meter.

Matching : Refers to adjustment of the effective impedance of a load with respect to that of the source, so as to ensure maximum power transfer.

Matching Transformer : Refers to a transformer which couples a source to a load, and designed to ensure maximum energy transfer in spite of the fact that the load impedance is different from the source impedance.

Maximum Demand : Refers to the highest value of the power, volt-amperes, or other quantity like the current taken within a demand assessment period (*e.g.* day, month or year). The appropriate quantity gets assessed generally by integration over each successive demand integration period (*e.g.*, half an hour).

Maximum Power Transfer Theorem : In order that a source of electromotive force E and internal impedance z shall be able to deliver maximum power to a load of impedance Z, then z and Z must be conjugate; *i.e.*, if z=r+jx then Z=R-jX with R=r and X=x. With the reactances equal and opposite in sign, there exists no resultant reactance in the circuit, and in this respect the current would be greatest because the total impedance has been r+R. If, further, R=r, then the load power would be the maximum possible.

Maxwell : Refers to the electromagnetic c.g.s. unit of magnetic flux. The flux in maxwells is commonly referred to as the number of 'lines'. A magnetic flux linking a single-turn coil and altering at the rate of 1 maxwell/s induces a voltage in the coil of 10^{-8}V. In terms of the weber, 1 maxwell=10^{-8} Wb.

Maxwell Bridge : A real-product a.c. bridge of the form which is shown in Fig. 3. The balance has been independent of the frequency.

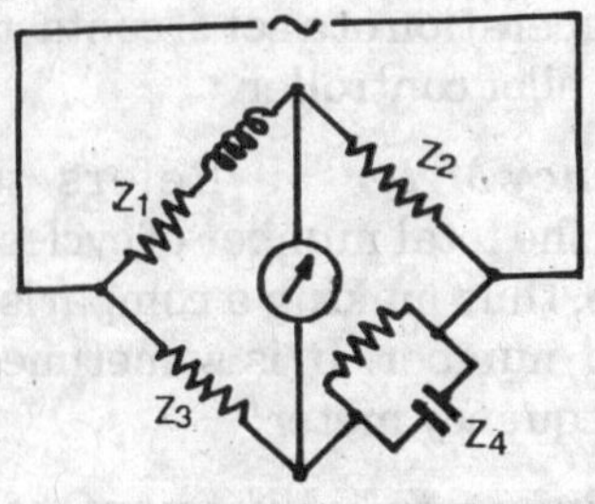

Fig. 3. Maxwell bridge circuit.

Maxwell's Laws : Refers to the basic laws of electro- magnetism which were formulated by Maxwell on the basis of the work of Gauss, Ampere and Faraday, and completed by his own concept of displacement current. The laws have been, in integral form:

1. The magnetomotive force, or line integral of the magnetic field strength H, round any closed path, would be equal to

   $\int_0 H.dl = I = I_c + I_d$

   It is termed as Ampere's law, with the provision that I is including both conduction current I_x and displacement current $I_d = \partial\psi/\partial t$, where ψ is the electric flux through the path.

2. The total electric flux emerging from a charge Q has been equal to Q. Then, the surface itegral over any closed surface of the outward electric flux desnity D would be given as follows:

   $\int_s D.ds = Q$

   If there has been no net charge enclosed within the surface, the integral has been zero. This is termed as Gauss's law.

3. Round any closed path encircling a magnetic flux Φ changing with time, there has been an induced voltage e which has been the line integral of the electric field intensity E.

   $\int_0 E.dl = e = -\partial\Phi/\partial t$

   It is called Faraday's law of electromagnetic induction.

4. Magnetic flux has been a solenoidal entity that is, it always comprises a structure of closed loops and there has been no 'source' of magnetic field. The surface itegral, over any closed surface in a magnetic field, of the flux density B must always be zero :

   $\int_s B.ds = 0$

Maxwell Theorem : Refers to a formalisation of Kirchhoff's mesh law which is used for the solution of networks. To each closed mesh in a network has been assigned a circulating current. The node law has been then automatically satisfied at each junction. Applying the mesh law to each closed mesh, a set of simultaneous equations could be written down for solution.

M.C.D. : Miniature Circuit-Breaker.

Mechanical Rectifier : Refers to a full-wave rectifier which incorporates a synchronously rotating or oscillating commutator.

Megger : A portable insulation-resistance tester which consists of a high-range ohmmeter and a hand-operated generator.

Mercury-arc Rectifier : Refers to a rectifier in which rectification occurs in gas conduction through mercury vapour at low pressure.

Mercury Motor Meter : Refers to an integrating meter which is embodying a motor, part of the moving portion being immersed in murcury.

Mercury Switch : Refers to the enclosed switch that provides continuity between two conductors by immersing them in a common mercury pool. The contact resistance has been lower than any obtainable by a pair of soild metal-to-metal surfaces.

Mercury-vapour Lamp : A discharge lamp having mercury vapour, useful light being emitted from or excited by the current through this vapour.

Merz-price Protection : Refers to a system of balanced current protection for power equipment and transmission circuits in which the ingoing and outgoing currents get compared by means of current transformers.

Merz Unit : Refers to a maxmimum-demand indicator unit for attachment to an integrating meter registering kWh or kVAh. A pointer gets moved forward by a spring-loaded mechanism which is operated by the meter register.

Mesh Connection : Refers to the arrangement of phase windings into a closed loop by the successive connection of the end of one to the beginning of the next, generally the windings give rise to (or have applied to them) voltages, and carry currents, differing only in time phase in a symmetrical manner. If the phase voltage of an m-phase mesh interconnection have been as follows :

$$V_1 = V\angle/0$$

$$V_2 = V\angle/-(2\pi/m)$$

$$V_m = V\angle\, l\,(m-1)\,(2\pi/m)$$

then the voltage AB in Fig. 4 is V_1, the voltage AC is V_1+V_2, etc. The current leaving at line A would be $I_m - L_1$, etc.

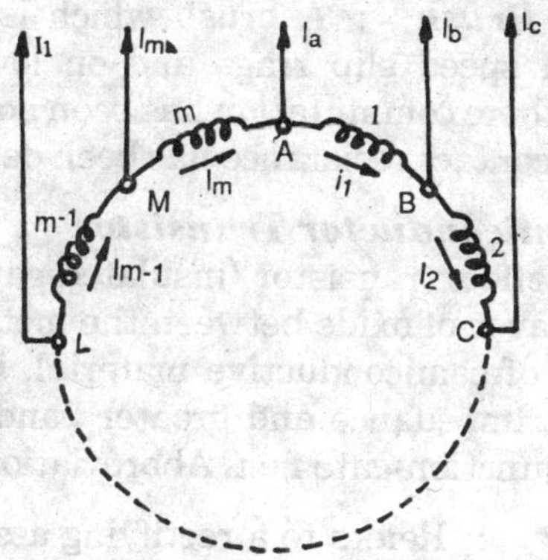

Fig. 4. Currents in a Mesh Connection

Mesh-current Analysis : The term used for a systematic method of solving for the currents in a network. It has been based on the Maxwell theorem, in which each mesh or closed loop of a network has been assigned a fictitious circuital current, and the assembly of all such currents conforms automatically to Kirchhoff's nodal law.

Mesh Voltage : Refers to the voltage between any two lines of a symmetrical three-phase system, or between two consecutive lines of a symmetrical six-phase system.

Metadyne Converter : A machine which is basically similar to a Metadyne generator with the supplementary set of brushes connected to an external d.c. supply so tht the output power does not need any appreciable mechanical power input to the trnasformer.

Metadyne Generator : A rotating amplifier of the cross-field excited type. It has been a quick-response d.c. generator, which takes in power at constant voltage through one set of brushes and delivers it at constant current through a second set.

Metal-clad Switchgear : Refers to a switchgear in which all live parts have been enclosed within a metal casing and normally insulated with oil or compound. The metal casing can get earthed.

Metal-enclosed Switchgear : A switchgear in which the whole equipment gets enclosed in a metal casing capable of being earthed.

Metal-graphite Brush : A brush which is used on low and medium speed slip rings and on low-voltage d.c. generators where commutation has been not difficult and where a low contact resistance has been desirable.

Metal-oxide Semiconductor Transistor : Refers to a type of field-effect transistor (insulated-gate f.e.t) which is having a layer of oxide between the metallic gate and the channel of semiconductive material. it is having a higher input impedance and greater bandwidth than a comparable junction-gate f.e.t. Abbreviation: m.o.s.t.

Metal Rectifier : Refers to a rectifying assembly of cells which consist of either :

(a) a semiconductor, such as selenium or cuprous oxide, in contact with a metal, or

(b) two semiconductors of different types, usually formed from a single piece of pure germanium or silicon by alloying or diffusing two different materials into opposite faces.

Meter : See integrating meter.

M.H.D. : See magnetohydrodynamic.

M.H.D. Generator : See magnetohydrodynamic generator.

Mho. : Obsolete term for the unit of conductance. It has been superseded by the siements (symbol : S).

Mica : It is a complex alumino-silicate of potassium, magnesium and iron, or of partial combinations of sodium lithium, titanium and vanadium. Micas are valuable heat-resistant electric insultants.

Mica Cone Of a commutator, this term refers to a V-shaped mica-compound ring which insulates a metal clamping ring from the bars. It is also termed as a mica V-ring.

Micanite : Refers to a flexible insulating material which is formed of mica splittings and bonded into a sheet by a flexible gum, bitumen or synthetic adhesive.

M.I.C.C. : Mineral Insulated Copper-covered (cable).

Microcircuit : Refers to a highly miniaturised assembly of circuit elements considered as a single unit.

Microelectronics : Deals with the techniques of making and applying microcircuits.

Microgap Switch : Refers to a switch for low-power low-voltage a.c. circuits in which the gap between the contacts when open has been if the order of 0.125 mm.

Microphone : A transducer which is used to convert acoustic vibrations to an analogous a.c. signal.

Microradiography : Refers to the examination of extremely thin sections of materials by using X-rays, the resulting microradiograph being subsequently optically magnified before examination.

Microtron : Obsolute term for electron cyclotron.

Microwaves : Radiowaves of frequencies higher than 1000 MHz.

Milker, Milking Booster : Preferably called a milking gen- erator.

Milking Generator : Low-voltage d.c. generator which is used for charging individual cells of a battery.

Millman Theorem : It is a network theorem that, in suitable cases, has been much more direct tan the application, for exampal, of Kirchhoff's laws. It is also termed as the parallel-generator theorem. It states that the common terminal voltage V of a number of sources connected in parallel would be as follows :

$$V=I_{sc}Z$$

where I_{sc} represents the sum of the short-circuit currents of the generators, and Z the parallel sum of the impendences between the common points. The theorem has been found to be applicable when all the generators are having the same frequency.

Miniature Circuit-breaker : Circuit-breaker rated at about 60 A or less, that may be used as protection in a domestic installation.

Miniature Edison Screw Cap : See Edison screw cap.

Minimum Fusing Current : Refers to the minimum current at which the fuse link in a fuse will operate in a specified time under specified conditions.

Minority Carrier : Charge carrier which forms only a small proportion of the total number of carrier in a 'doped' semiconductor. In p-type material the minority carriers have been electrons; in n-type material they are holes.

Minus Tapping : Refers to a tapping on a winding, having its position such that fewer turns have been included in the active part of the winding than are required for the service voltage or current ratio.

Mirror Galvanometer : A Galvanometer which is having a mirror attached to the moving part, and reflects a beam of light on to a scale or reflects the image of the scale into a telescope.

Misfire : Refers to the failure of ignition in a gas or vapour-filled electronic tube or valve in periodic operation as a static converter.

M.M.F. : See magnetomotive force.

Modulation : Refers to the process whereby a characteristic of a carrier (or propagated) wave of the high angular frequency gets varied in accordance with the time variation of the intelligence to the transmitted.

Moissan Arc Furnace : Refers to an indirect-arc furnace in which an arc gets struck between two carbon electrodes placed above the material to be heated (see Fig. 5). The effect is therefore entirely thermal, without electrolysis.

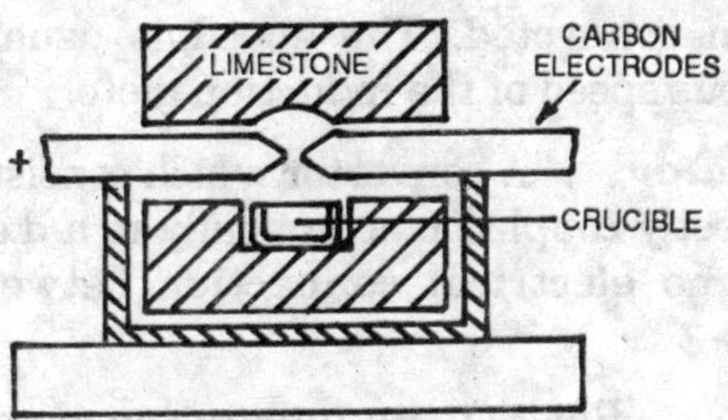

Fig. 5. Principle of the Moissan indirect-arc furnace

Mollerhoj Cable : Refers to a pressure cable in which the cores have been laid side by side in flat formation (Fig. 6). The lead sheath get supported by circumferential metal tapes together with corrugated metal strips bound on to the flat faces of the sheath by copper wires. All oil of low viscosity finds use for the impregnating medium.

Fig. 6. Mollerhoj Cable.

Monostable Relay : Refers to a relay with a single non-energised rest condition to which it automatically returns on removal of an energisation.

M.O.S.T. : Metal Oxide Semioxide Semiconductor Transistor.

Motor : Refers to a machine which is used for coverting electric energy into mechanical. A current-carrying conductor in a magnetic field gets acted on by a force proportional to the current and the field strength. This principle has been applied to the rotation of wound rotor.

Motor Converter : A converter which consists of a wound rotor induction motor coupled to a synchronous converter. The energy transmission has been partly mechanical and party electrical, as the two armatures and their circles have been connected. The speed is usually half the synchronous speed of the induction motor.

Motor Generator : A convertor which consists of any a.c. motor directly coupled machanically to a d.c. generator. There is no electrical connection between the two machines.

Motor Meter : Refers to an integrating meter which embodies a motor of some type.

Moving-coil Galvanometer : See galvanometer.

Moving-coil Instrument : It is the commonest type of indicating instrument, although by itself it measures only direct current. It consists of a fixed permanent magnet, in the air-gap of which is a coil carrying current, which, combined with the steady magnetic flux, produces motion, shown by a pointer (see Fig. 7).

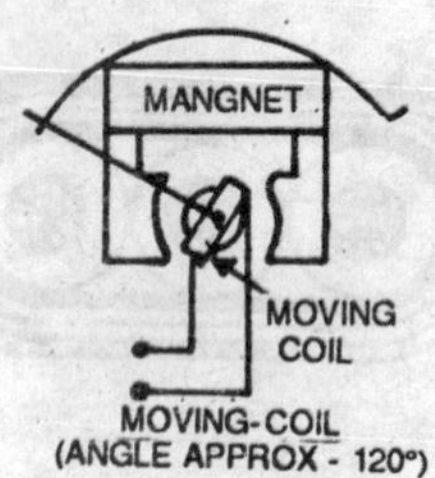

Fig. 7. Moving-coil Instrument

Moving-coil Regulator : Refers to a form of voltage regulator. An auto-transformer having straight limbs has been provided with two similar windings connected in opposition, so that the flux through each follows the path indicated in Fig. 8. Over these windings moves a short-circuited coil, if this coil had not been present the voltage across each winding would be the same, and the point A, would be midway in potential between the two supply terminals. If the coil gets moved to a position covering the upper winding, the magnetomotive force due to the currents in the short-circuited coil will oppose that of the upper winding, thereby reducing the upper flex almost to zero. The potential of A would therefore be almost that of the upper supply terminal and the output voltage be nearly equal to the supply voltage.

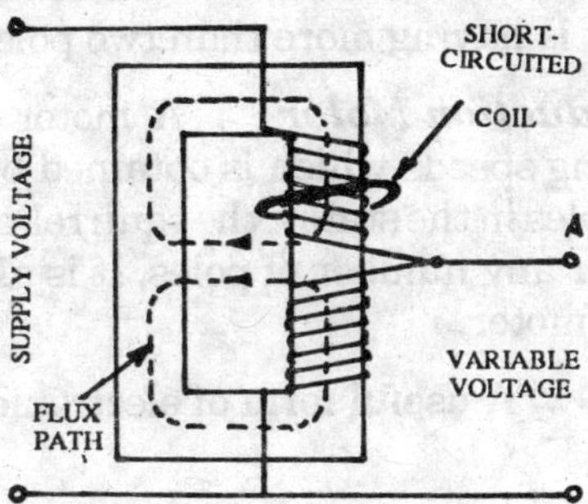

Fig. 8. Moving-coil Regulator.

Moving-coil Relay : A relay which consists of a permanent magnet and coil, the latter mounted to move in the permanent magnet flux against spring restraint, and to operate the relay contacts.

Moving-iron Instrument : It is a common form of indicating instrument. A fixed coil and a moving iron could be arranged for mutual attraction; alternatively, two irons may be similarly magnetised by the same coil and mutually repel each other; if now one iron gets fixed and the other pivoted the movement will depend on the current. Both these attraction and repulsion types have been in use. Moving-iron instruments read voltage and current equally will on direct current and alternating current of power frequencies.

Moving-magnet Instrument : Refers to a moving-iron instrument in which the movable component is having a piece of permanently magnetised material.

Multi-break Circuit-breaker, Switch : Refers to a circuit breaker or switch in which there have been two or more breaks in series in each pole or phase.

Multiple Earthing : See protective multiple earthing.

Multiple Tariff : A tariff in which different rates have been charged for units supplied for various purposes.

Multi-point Heater : Water heater which supplies hot water to a number of outlets. It may be a pressure heater or a cistern heater.

Mutli-polar Machine : Refers to a machine in which the field magnet is having more than two poles.

Multi-speed Induction Motor : A motor having two or four operating speeds which is obtained by changing the number of poles in the stator, the squirrel-cage rotor being adaptable for any number of poles. It is also termed as a change-pole motor.

Multivibrator : A useful form of electronic relation oscillator.

Mumetal : A ferromagnetic alloy, approximately 75% nickel, 5% copper and 2% chromium. Its important properties have been high permeability and low hysteresis loss in relatively low magnetising fields. The initial and maximum relative permeabilities have been of the order of 30,000 and 100,000.

Murray Loop Test : Refers to a cable fault localisation test. The faulty conductor, which must be continuous, gets looped to a sound conductor of the same length and size, and a resistance bridge and galvanometer get connected across the open ends of the loop (Fig. 9). A battery or other d.c. source gets connected between the bridge tap and earth. A Wheatstone brdige arrangement would be formed by the circuit, two of the arms being given by the resistances a and b, and the other two by the parts of the cable loop on either side of the fault. Balance would be obtained by adjustment of a and b, and the distance to the fault from the test end is given by

$$x=1.2a/(a+b)$$

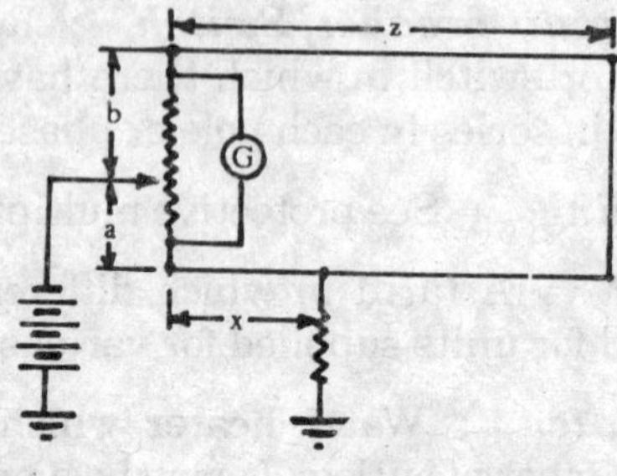

Fig. 9. Murray Loop Test for locating a fault in a cable.

where a represents the arm connected to the faulty core and 1 represents the route length of the cable. Also Werren overlap test, Fisher loop test.

Mush Winding : Refers to the winding for small a.c. machines in which the conductors have been dropped individually into lined slots, the end connections then being insulated separately. It is also termed as a random winding.

Mutual Coupling : Refers to the coupling of coupled circuits in which the two circuits are having a partly common magnetic circuit.

Mutual Impedance : Alternative name for transfer impedance.

Mutual Inductance : Refers to the ratio of the magnetic flux-linkage in an electric circuit to the current flowing in another circuit linked magnetically with the first. The two circuits would be having unit mutual inductance of one henry if (a) the energy stored in the common magnetic field has been 1 J when the current in each circuit has been I A. (b) a primary current changing at the rate of 1 A/s induces a voltage of 1 V in the secondary, or (c) the number of flux-linkages with the secondary totals 1 Wb-turn when the primary current is 1 A.

Mutual Surge Impedance : Refers to the quotient of the r.m.s. voltage of a travelling wave in one transmission line by the r.m.s. current of the resultant travelling wave in another transmission line system.

M.V. : See medium voltage.

Natural Frequency : Refers to the frequency at which an oscillatory system will oscillate when provided with energy and then left free of restraint.

Natural Graphite Brush : A brush which is similar to a carbon brush but with natural graphite (*i.e.*plumbago) instead of carbon. The excellent lubricating properties of graphite and its high thermal conductivity have made these brushes suitable for the larger heavy-duty d.c. generators and motors and for high-speed applications like turbo-alternator slip rings.

Natural Load : Refers to a load impedance which is normally nearly resistive and equal to the characteristic impedance of the transmission line that it terminates. The term finds use in connection with energy transmission over long power lines. Telecommunication lines always get terminated on their natural load so as to avoid signal reflection. They are said to be matched.

Needle Gap : Refers to a spark gap which is used for measuring voltages of a few kilovolts. It is having a larger and therefore more convenient, spacing than a sphere gap at these low voltages. Consistent rules have been however, difficult to obtain.

Negative Booster : A booster which gets arranged to reduce the voltage supplied from another source.

Negative Feedback : Means return of energy from the output of an equipment to the input having reverse polarity or in anti-phase. It will be able to reduce the gain of an amplifier but neutralises a large amount of distortion. It is also termed as degenerative feedback.

Negative Feeder : In an electric-traction system this term refers to the feeder connecting the track rails or negative conductor rail to the negative busbars at the substation or generating station. It is also termed as a return feeder.

Nagative Feeder-booster : Refers to a booster which is arranged to reduce the potential difference between two points of an earth return.

Negative Glow : Refers to the luminous glow in a discharge tube that may take place round the cathode at low gas pressures.

Negative Phase Sequence : Refers to a sequence of phase voltages or current which has been opposite to the normal (or positive sequence) in a polyphase a.c. system. If the positive sequence of three phases has been ABC, the ACB would denote a negative sequence.

Negative Resistance : The term used to denote the property of a circuit parameter such that the voltage across it is $V = -IR$.

Neon Tube : Refers to an evacuated glass tube or bulb, having two electrodes in an atmosphere of neon or other gas at low-pressure.

Neoprene : Trade name for the first commercial synthetic rubber, polychlorobutadiene which gets derived from acetylene and hydrochloric acid. It finds use as a cable sheath.

Neper : Refers to the dimension less unit of attenuation factor.

Network : A complex circuit consisting of a number of branches, which are connected at junctions or nodes, and forming closed loops or meshes.

Network Analyser : A device which facilities the calculation of the performance of electrical power systems and other electric circuits. Network analysers constitute a distinct group of electric circuit machines in the general

classification of analogue calculators. Arrays of circuit units denote the parameters of generators, lines, transformers and loads of a system.

Network Analysis : Refers to the determination of the voltage, current, power dissipation,energy storage, etc., in an electric circuit. All methods of analysis has been based on Kirchhoff's laws which, being based on the continuity of current and the principle of energy conservation, have been always applicable.

Network Function : See system function.

Network Synthesis : The term used for the process of designing an electric network to have specified properties; a process which is the inverse of network analysis. The properties required of the network to get designed have been expressed by a system function which describes what kind of response has been to result from a given stimulus.

Neutral : Refers to a point, conductor or region in an electrical system or device having symmetry in some special defined sense, *e.g.* the position of the brush axis of a machine such that the electromotive force appearing at the brush has been a maximum or minimum or a point on a supply system which is normally having the same magnitude of potential difference from each of the line conductors. In the latter case the point has been usually at earth potential.

Neutral Conductor : A conductor which is connected to the neutral point of a supply system and laid along with the supply conductors. The neutral points of load equipment could be connected to it. It is generally earthed at the supply end of the system. The use of a neutral conductor provides a choice of voltage for supplying a two-terminal load.

Neutral Inversion : A phenomenon causing the phase-to-neutral voltages of a supply system becoming unequal so that the neutral-point potential gets moved away from the geometrical centre of the supply-voltage triangle, and takes a position possible outside the triangle (Fig. 1) where all phase-to-neutral voltages have been far above their normal values. Such a situation may arise if the

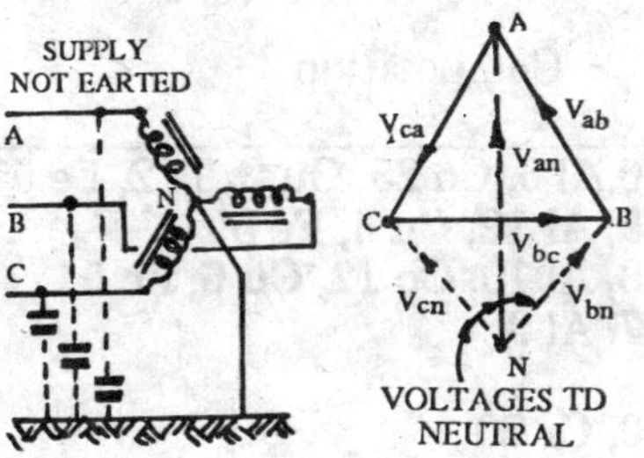

Fig. 1. Neutral Inversion.

impedances to neutral of the three phase conductors get widely unbalanced; this could take place in the normally balanced circuit where a cpacitance has been in parallel with an iron-cored inductance such as that of an unloaded power or voltage transformer, the only earth point on the system being that shown in the figure.

Neutral Position : Refers to a position of brushes in an electric machine. In an a.c. machine it is that position which is obtained when the axes of the main stator and rotor windings coincide; in a commutator machine, it has been the position that provides equal speeds in either direction of rotation when the machine is run at a constant load.

Neutral Zone : Refers to the part of the commutator of a d.c. machine in which, when the machine has been running at no load, the voltage between adjacent bars has been almost zero.

Newton : It is the SI unit of mechanical force (symbol: N). It may be defined as that force which imparts to unit mass (1 kg) a unit acceleration (1 m/s^2). It is also that force which, expended over the distance of one meter, does work amounting to one joule, so relating it directly to the usual electrical units.

Nickel Alloys : Mixtures of nickel with other elements, particularly iron.

Nickel alloys

Alloy	Composition	Application
Alcomax	Ni 10,Al 6, Co 25, Cu 2, Ti 2, Fe 55	
Alni	Ni 25, Al 12, Cu 4, Fe 59	Permanent
Alnico	Ni 18, Al 10, Co 12, Cu 6, Fe 54	magnets
Alumel	Ni 97, Al 3	Thermo-Couples
Brightray	Ni 80, Cr 20	Heating elements
Chromel	Ni 80, Cr 20: ect.	Thermo-couples
Constantan	Ni 40, Cu 60	Resistors
Hypernik	Ni 50, Fe 50	High-μ
Mumetal	Ni 76, Cu 5, Cr 2	High-μ
Nichrome	Ni 65, Cr 15, Fe 20	Heating elements
Permalloy	Ni 78.5, Fe 21.5	High-μ
Permalloy C	Ni 78.5, Mo 3.5 Fe 18	High-μ
Supermalloy	Ni 97, Mo 5, Mn 0.5, Fe 15	High-μ

Nickel-iron alloys with 45-80% nickel are having high permeability and low loss at small magnetic field intensities. Excellent magnetic properties have been also obtained from alloys of nickel, iron and additional metals. The properties of the three permanent-magnet materials in the main table have been (rfemanence B*r* in Wb/m^2, coercivity H_c in A t/m, and $(BH)_{max}$ J/m^3) :

	B*r*	H*c*	(BH)*max*
Alcomax	1.2	40,000	30,000
Alni	0.6	40,000	10,000
Alnico	0.7	44,000	13,000

The high-permeability (high-μ) materials listed are having initial and maximum relative permeabilities, coercivities in At/m, saturation densities B*s* in Wb/m^2, and resistivities ρ in μΩ-cm, as follows :

	μ_i	μ_{max}	H_c	B_S	ρ
Hypernik	8,000	60	8	1.60	45
Mumetal	20,000	80	2.5	0.75	42
Permalloy	10,000	100	4	1.07	17
Permalloy C	21,000	75	4	0.85	55
Supermalloy	100,000	1000	0.3	0.80	65

Nickel-Cadmium Cell : Refers to a steel-alkaline cell in which the electrolyte has been dilute potassium hydroxide, the positive plate nickel hydrate, and the negative plate cadmium having a small proportion of iron. The use of cadmium provides this cell a lower charging voltage than the nickel-iron cell and a reduced internal resistance, so that for those duties where the difference between charge and discharge voltage is important or where the ability to give high rates of discharge with a well-maintained voltage has been essential, the nickel-cadmium cell has been in more general use. It is used for such duties as engine statting, switch closing and tripping, starting and lighting purposes on vehicles, for lighting and manoeuvring on trolleybuses, etc.

Nickel-iron Cell : It is a steel-alkaline cell in which the electrolyte has been dilute potassium hydroxide, the positive plate nickel hydrate, and the negative plate iron. Its use has been principally confined to such duties as traction, house-lighting, etc.

No-lag Motor : A compensated induction motor which is having stator and rotor similar to those of a Schrage motor but with the brush gear fixed.

Node :

1. In a standing wave the term refers to the appearance of a minimum value.
2. In a network the term refers to the end-terminal of a branch or a point where two or more branches are joined.

Node-voltage Analysis : Refers to a systematic method of solving for the voltages in a network. Each junction point has been a node. One node has been used as a

reference so that the voltages of all the other independent nodes have been found with respect to it. All sources have been converted to equivalent current generators, and the total currents fed into the nodes have been expressed by the simultaneous equations

$$I_1=V_1Y_{11}+V_2Y_{12}+......+V_nY_{1n}$$

$$I_2=V_1Y_{21}+V_2Y_{22}+......+V_nY_{2n}$$

....

$$I_n=V_1Y_{n1}+V_2Y_{n2}+......+V_nY_{nn}$$

Here $Y_{11}, Y_{22}.....Y_{nn}$ denotes the self-admittances of the nodes, *i.e.*,the sum of all the admittances of the branches metting at nodes 1, 2......n. Terms of the form Y_{12}, Y_{1n}, Y_{3n}......denotes the mutual admittances respectively between the nodes 1 and 2, 1 and n, 2 and n......In each case $Y_{pq}=Y_{2pq}$ and if all the voltages of the independent nodes are regarded as having the same polarity with respect to the reference node, then all the mutual admittances would be taken as negative.

Noise : Unwanted electric signals and their audible result. The term includes transformer noise, which is audible and is having its origins in magnetostriction and in magnitic forces on the core laminations, and circuit noise, which has been random and may occur to electron movement (thermal-agitation noise).

Nominal T and II Networks : The term used for the approximate representations of a transmission line, for the purposes of analysis. The line can be regarded to tbe passsive quadripole of T or II formation. Such a quadripole can be devised to provide characteristics which are identical with those of the line for any given frequency. Such T and II sections would then be termed as equivalent.

Non-bleeding Cable : Refers to a pre-impregnated or drained impregnated cable. Under working conditions there has been no danger of the impregnating material exuding.

Non-pressure Heater : A water heater which is having an open outlet. It is connected to a cold-water supply and is controlled by a stop valve on the inlet side. It is also called a displacement heater.

Non-reative Load a load in which the current and voltage have been in phase at the terminals.

Normal Bend : Refers to a conduit fitting which is having a longer radius than an elbow. It is used to connect two lengths of conduit at right-angles. A similar fitting for connecting at $1\frac{1}{2}$ right angles has been a half-normal bend.

Norton Theorem : See Helmholtz-Norton theorem.

Nose Suspension : Refers to a method of mounting a traction motor on an electric truck. One side of the motor gets supported on the axle by using suspension bearings and the other on the framework by using a lug projecting from the motor case.

Notching Controller : A switch which is actuated mechanically by the mehanism of a tap changer to ensure that the tap changer, having once been set in motion, continues its movement until the change in tapping has been complete. It is also termed as a sequence switch.

N-p-n Transistor : A transistor which consists basically of two outer n regions separated by a thin p layer. The operation has been similar to that of a p-n-p transistor, except that electrons flow through the p-type base instead of holes through an n-type region. The main difference, practically, has been that the bias potentials are having a polarity opposite to that for the p-n-p type (*i.e.* positive on the collector and negative on the emitter).

N-type Semiconductor : A quadrivalent semiconductor like germanium, which includes a small, accurately controlled amount of a pentavalent donor element so that the cystal lattice is having free electrons to act as charge carriers.

Nuclear Energy : The energy released in nuclear reactions or transitions.

Nuclear Fission : A particular type of nuclear reaction in which a heavy nucleus is split into two nuclei of lower mass. In each nuclear fission a large amount of energy gets released and most appears as kinetic energy of the fission fragments. More than one neutron is emited per fission. Hence a chain-reaction can be initiated by these emitted neutrons. The process is initiated and controlled in a nuclear reactor.

Nuclear Fusion : Refers to the combination of fusion of light elements into heavier with the release of energy. This may be attained in a thermonuclear reactor or plasma.

Nuclear Power Station : A power station which produces electric energy from nuclear energy.

Nuclear Reactor : A plant which is used for the controlled production of a neutran chain-reaction producing nuclear fission. The most useful fuel for a nuclear reactor has been natural uranium, comprising 0.7% fissile uranium-235, the rest being uranium-238. Uranium-235 is the only fissile material occurring in nature, but two other fissile materials, plutonium and uranium-233, can be made artifically by neutron bombardment of uranium-238 and thorium respectively. A nuclear reactor is sometimes termed as an atomic pile.

Null Detector : Refers to an instrument, such as a galvanometer, in a bridge which is used to indicate that equality of values gets established between as unknown and a known variable quantity.

Null Measurement : Refers to a method of making an electric measurement in which the quantity to be measured, which must be steady, has been compared with a known quantity using an indicating instrument as a null detector.

Nylon : A thermoplastic material which is mainly used in the electrical industry as cable sheathing, and for the production of small moulded components. It is having great mechanical strength.

Nyquist Criterion : Refers to a means whereby the degree of stability or instability of a feedback system could be assessed from its complexor locus diagram.

O.C.B. : Oil Circuit Breaker.

O-class Insulation : Refers to a class of insulating materials now designated Y-class insulation.

Oersted : Refer to the c.g.s. unit of magnetetic field stregth. It is equal to 4π/10 times the ampere turns per centimetre.

Off-peak : During a period when the demand on the power supply system has been not at a maximum *e.g.* at night.

Ohm : Refers to the SI unit of resistance (symbol: Ω). A conducting path is having a resistance of 1Ω when the passage of a current of 1 A needs a potiential difference across the path of 1 V.

Ohmic resistance : Refers to the resistance which is offered by a conductor to the free flow of a direct current and undisturbed by such influences as eddy currents, thermo e.m.f.s. or pinch, skin and proximity effects.

Ohmmeter : An instrument having voltage and current coils to read the quotient of the voltage across an unknown resistance and the current throught it. A major application, incorporated in an instrument having a hand-driven magneto generator, has been the measurement of insulation resistance.

Ohm's Law : According to this law, the current I in a conductor is proportional to the potential difference V

across its ends. The usual form of the law has been V/I=R, where R is the resistance (unit, ohm).

Oil Circuit-breaker : Refers to a circuit-breaker in which the arc has been drawn in oil and its high temperature makes the formation of a gas bubble consisting mainly of hydrogen. The turbulence of the gas and the cooling effect of the surrounding oil tend to disperse and de-ionise the arc path at a current zero and restore its dielectric strength. Arc-control devices have been used to confine the arc and improve performance. The main types of oil-circuit-breaker include the bulk-oil circuit-breaker and the small-oil-volume circuit-breaker. Impulse circuit-breakers belong to the second category.

Oil Conservator : A chamber which is fitted above the main tank of an oil-filled transformer. Its capacity is that of about 10% of that of the main tank. It gets connected to the tank by a small-diameter pipe so that oil may expand freely into it, and gets intended to ensure that the oil has been maintained in good condition, firstly by excluding moisture and secondly by reducing the area of the oil in contact with the air.

Oil Expansion Chamber : See oil conservator.

Oil-filled Cable : A cable design including metal spirals in the interstices of a three-core (Fig. 1) or in the centre of the strand of a single-core cable. The dielectric gets impregnated with a thin oil after a reinforced-lead or aluminium sheath is applied. As the temperature of the

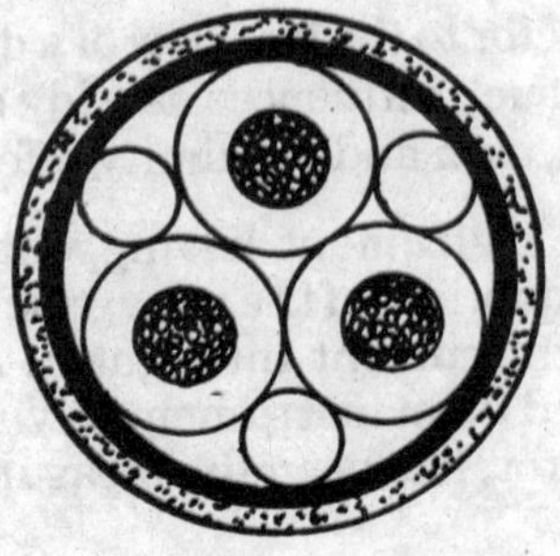

Fig. 1. Oil-filled three-core cable.

cable gets increased by loading, the excess oil caused by volumetric expansion passes through the metal spirals into membraneous tanks which are situated along the route. When the cable cools the oil gets forced back into the cable by the movement of the flexible membrances in the tanks. Oil-fitted cables have been used at 33 kV and about.

Oil Switch : A switch in which the contacts have been oil immersed. It has been commonly applied to motor control.

Oil-tank Fuse : Refers to a fuse in which the fuse-link has been totally immersed in an oil tank. The oil acts as an arc-extinguishing medium.

Oilostatic Cable : Oil-filled cable in which the sheath has been eliminated and the three insulated conductors have been pulled into a steel pipe which has been then filled with oil maintained at high pressure.

On-line Working : Refers to a mode of operation of a unit of an automatic data processing system. The unit has been connected to the main part of the system and operates without significant delay when the output from another unit becomes available.

On-load Tap Changing : The term used for varying the effective number of turns in the primary or secondary windings of a transformer when the transformer is on-

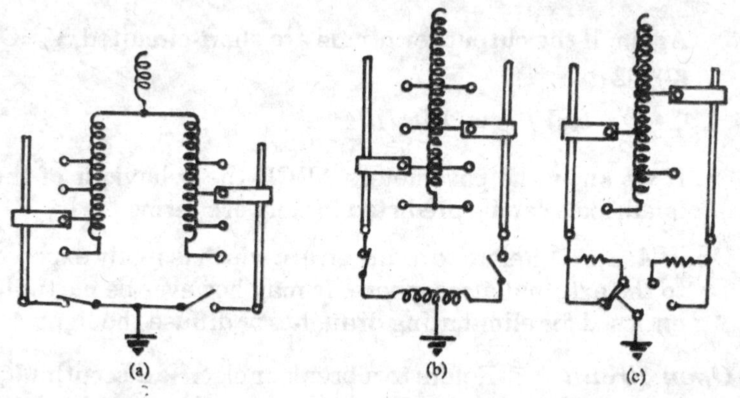

Fig. 2. Three methods of tap-changing, involving : (a) parallel winding, (b) an inductor bridge, and (c) a resistor bridge.

load. There must occur no interruption in the winding circuit while the selector switches operate to change the active tappings. The methods in use are illustrated in Fig. 2.

One-port, one-port Network : A network having a single port.

Open and Closed Test : Refers to a modification of the Murray loop test for locating faults in cables. This test is designed for use when the insulation of the return conductor has a low value.

Open-and Short-circuit Characteristics : Refers to the relations from which all the essential details of the circuit behaviour of an a.c. device can be in principle, derived. Suppose a passive quadripole is considered in which input and output voltage and current, V_1I_1 and V_2I_2 are related by the general parameters ABCD so that

$$V_1=V_2A+I_2B$$

and $$I_1=V_2C+I_2D$$

If the receiving end has been open-circuited, then $I_2=O$, whence $V_1=V_2A$ and $I_1=V_2C$. Thus, if the input voltage and current, and the output voltage, have been measured in magnitude and phase, then

$$A=V_1/V_2 \text{ and } C=I_1/V_2$$

Again, if the output terminals are short-circuited, $V_2=O$, giving

$$B=V_1/I_2 \text{ and } D=I_1/I_2$$

If we know the parameters ABCD the behaviour of the quadripole can be predicted in network terms.

Open Arc : Refers to a carbon arc which is freely exposed to the external atmosphere. It may however be partially enclosed for eliminating draughts or diffuse the light.

Open-circuit : Refers to a break or electric discontinuity in a circuit through which current could normally flow. The term is applied to a generator o. transformer.

Open-type Switchgear : Refers to a switchgear in which the various items have been mounted on steel or concrete structures and connected with bare conductors. The use of air as the main insulation reduces the cost.

Optical Fibre : Refers to a hair thin fibre of glass or plastic material having a high value of refractive index, clad in a very thin sheath of a material having a lower value of refractive index. It constitutes a flexible pipe which is capable of guiding light over substantial distances and also round bends, and a bundle with orderly arranged fibres provides a channel for a coherent transmission of an image which could be employed be used to inspect cavities in industrial objects or in medical diagnostics. High-quality glass fibre cables are used in telecommunication and data transmission systems for guiding modulated light beams originating from lasers. Optical-fibre cables, combined with a light-emitting diode or a laser as the light source to produce modulated light signals, find use in industrial control systems.

Optical Pyrometer : A pyrometer which is able to compare the intensity of visible radiation from a body with that of the same colour and wavelength derived from a calibrated comparison source, usually a filament lamp.

Optoelectronic Semiconductor Device : Refers to a semi conductor device that either responds to electromagnetic radiation in the spectral region from ultra-violet to infra-red, or emits or modifies such radiation.

Orbital Accelerator : Refers to a direct particle accelerator which is having either a constant magnetic field in which the particles move in orbits having a series of arcs of circles of discrete and increasing radii (cyclotron, synchrocyclotron or Microtron), or a changing magnetic field having nearly constant orbit radius (betratron, synchrotron).

Oscillating Neutral : Refers to an insulated star-point which oscillates about the neutral potential at third-harmonic frequency. It may take place when the load comprises iron-core devices, and the insulated star-point of the generator is earthed.

Oscillograph : Refers to an instrument which is used for writing or recording oscillations. The term is sometimes uses for an oscilloscope. A mechanical form has been the Duddell oscillogrph.

Oscilloscope : An instrument which is used to represent oscillations in a visible form. Its most common form incorporates a cathode-ray tube.

Output Coefficient : A coefficient which expresses the volt-ampere rating of an electric machine per unit volume and per unit speed.

Overcompensation : Refers to the result of a compensating winding which is having a greater effect than the armature reaction which it opposes. A cumulative action would be obtained, tending to make output voltage rise as load current gets increased.

Overcompounded : Term referred to a compound-wound generator in which the series winding has been so designed that the voltage increases with the load.

Overcurrent Protection : Refers to the protection of a system against excessive currents. The fuse forms the basis of most small, simple distribution-system protection, combining overcurrent protection and fault isolation. Another basis for the over-current protection is the definite-minimum-time relay.

Overhang : The part of a winding is also termed as the end winding.

Overhead Distribution : Refers to the supply of power from generating stations or main substations by using tower-supported transmission lines to load centres or secondary substations (primary distribution), or the supply of power by similar means from secondary substations to consumers' premises (secondary distribution).

Overhead Line : Cables used in overhead distribution of power. In the early years conductors of cable were invariably of hard-drawn (h.d) copper used at moderate voltages with short span wood-pole supports. Alternative and stronger conductor materials were introduced for this purpose: cadmium-copper alloy, steel-cored copper, copper-cored steel, and finally aluminium alloy and steel-cored aluminium.

Overlap Angle : Refers to the time angle during which successive anodes are able to conduct simultaneously in a multi- anode rectifier.

Overlap Span : Alternating name for section gap.

Overload : Refers to a load on a machine or system which has been greater than it has been designed to withstand continuosly. It is expressed numerically as the amount in excess of the rated load.

Overload Protection : See overcurrent protection, overload relay.

Overload Relay : A relay which has been designed primarily to protect a motor or other equipment from damage due to currents in excess of normal. The relays can be divided into two main groups, thermal overload relays and magnetic overload relays.

Overvoltage : Refers to an increase in system voltage above the normal or declared values. It may be transient or sustained. A transient overvoltage may damage the system through electric breakdown but is not likely to cause breakdown through excess current resulting from the overvoltage. A sustained overvoltage may of course cause breakdown in the system through the excess current causing overheating.

Overvoltage Test : Test applied to transformers, etc., to prove the quality of the internal insulation. It could be an induced-voltage test or a separate source test.

Oyster Fitting : A robust light fitting which has been designed for attachment to bulkheads, etc., where space has been restricted, and emitting light simultaneously on both sides of the bulkhead.

Palmer Limit Switch : A special switch which is used with d.c. series motors on cranes, where the motor may generally be running at over double normal speed and quick stopping is necessary.

Pantograph : Refers to a sliding current-collectror which is used in electric traction with overhead wires. It has a low-shaped contact strip mounted on a lozenge-shaped or corss-arm frame which gets hinged to provide vertical motion for the contact strip.

Paper : Cellulose fibres which is felted to form a mechanically strong sheet. Cable papers are having a density normally from 0.7 to 0.9 g/cm^3 dry. To improve the dielectric strength of a cable paper it has been impregnated with a suitable insulating oil so that approximately one half of the cable dielectric has been occupied by the impregnant. The permittivity is approximately 2.25. Capacitor paper has been generally a kraft paper of the order of 10 to 15 μm thick which is impregnated with mineral oil or petroleum jelly. The effective permittivity at 50 Hz of an oil-impregnated paper has been of the order of 4.0. The loss angle at 50 Hz has been 0.0015 to 0.003; the insulation resistance at 20°C has been 2×10^4 ohm-farads; the approximate working stress has been 15 to 20 r.m.s. V/μm; and the working temperature range up to 90°C.

Paraffin Wax : A wax-like substance which is left after the distillation of petrolium. It is used for the impregnation of some capacitors, radio coils, etc. where the operating temperatures have been not high. It has a melting point of 50-60°C, resistivity at 20°C of 10^9 - 10^{13} MΩ per cm cube, and permittivity of 2.2.

Parallel-generator Theorem : Alternative name for the Millman theorem.

Parallel Operation : The term used for the operation of two or more machines or transformers with their terminals of similar polarity connected together. It means equality of terminal voltage. This condition puts restrictions on the characteristics of the machines, and affects stability and load sharing.

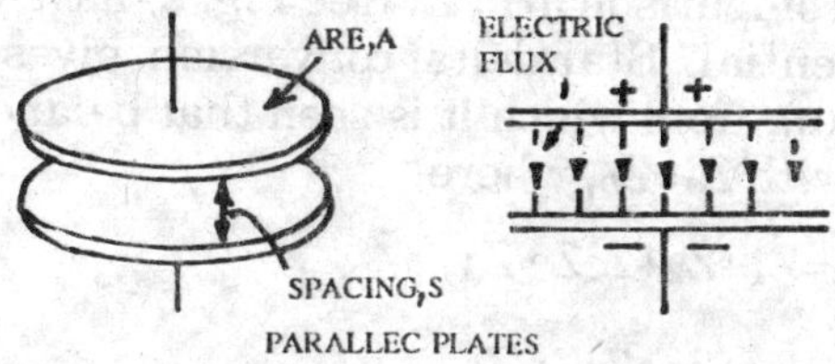

Fig. 1. A basic parallel plate capacitor.

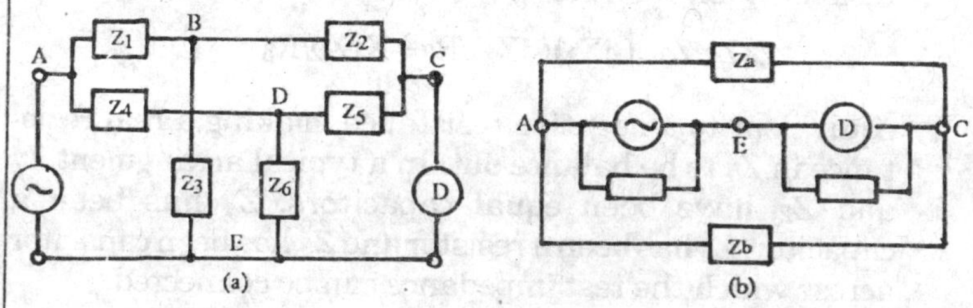

Fig. 2. A parallel-T bridge and its star/delta conversion.

The conditions that must be observed for the parallel connection of transformers on both primary and secondary sides have been that they should have the same voltage-ratio, the same polarity, the same phase-sequence, and zero relative phase displacement.

Parallel-plate Capacitor : A capacitor which consists basically of two equal parallel plates separated by a dielectric (Fig. 1). If the area of each plate has been A, their separation s, and the charge IC, then neglecating edge effects, the dielectric flux density at any point between the plates would be $D=1/A$, so that the electric field intensity $E=D/\in -1/A\in$. As the field is uniform, integration of E yields $Es=V=s/A\in$. Hence the capacitance would be as follows :

$$C=1/V=\in A/s \text{ farads}$$

Parallel-resonant Circuit : Refers to a resonant circuit having two parallel paths, of which one is having capacitance and the other inductance.

Parallel-T bridge : A six-arm a.c. brdige which finds use at radio frequencies. It has the advantage that one end of the supply, the detector, unknown impedance Z_3 and standard comparison arm Z_6 (see Fig. 2) have been all at earth potential. Star/delta conversion gives the form shown in (b), from which it is seen that balance could be obtained with $Z_a=Z_b$, where

$$Z_a=Z_1+Z_2+Z_1Z_2/Z_3$$

$$Z_b=Z_4+Z_5+Z_4Z_5/Z_6$$

If Z_1 and Z_2 are pure reactances of like sign and Z_3 is a resistor: then we have

$$Z_1Z_2/Z_3=(-jX_1)(-jX_2)/R_3=-X_1X_2/R_3$$

equivalent to a negative resistance, making a real resistance in Z_b to be balance out. In a typical arrangment Z_1 and Z_2 have been equal capacitors, Z_4 has been a capacitor, Z_5 has been a resistor and Z_3 has been capacitor across which the test impedance can be connected.

Paramagnetism : A phenomenon which is exhibited by materials where the relative permeability is greater than unity.

Parametric Amplifier : Refers to a reactance amplifier which works on the principle that if, an oscillatory LC circuit, either the capacitance C or the inductance L gets varied periodically at appropriate frequency and in appropriate phase, electric oscillations could be maintained in the circuit.

Particle Accelerator : A machine which is used for accelerating electrons or ions to high energies.

Accelerators could be divided into two main classes, direct or indirect, depending on whether the particles are accelerated directly by the direct voltage applied between two electrodes, or whether the acceleration has been obtained with the aid of an r.f. electric field. An indirect machine may be either a linear accelerator or an orbital accelerator. Examples of direct machines include the Cockroft-Walton multiplier and the Van de Graaff generator. Orbital machines include the cyclotron, synchrocyclotron, microtron, betatron and synchrotron.

Passive Circuit Element : Refers to a circuit element that in operation either consumes energy like a resistor, or may store energy without consuming any like a loss-free capacitor or inductor. A passive circuit or network is having only passive circuit elements.

Pasted Plate : Refers to a plate of a lead-acid cell which is made mechanically by the application of lead-oxide paste on to a grid. It is also termed as a Faure plate.

Paxolin : Trade name for a type of synthetic-resin-bonded paper.

P.B.F. : Paper-braided jute (insulated cable).

P.D. : Potential difference.

Peak Factor : Refers to the ratio of the peak value of an alternating or pulsating wave to its root-mean-square value. A sine wave is having a peak factor of $\sqrt{2}$.

Peak Load : Refers to the maximum demand on a paper supply system.

Peak Value : The greatest instantaneous value of a time-dependent quantity within a given time interval which is equal to one period for a periodic quantity.

Peak-to-valley (p-v) Value : Refers to the difference between peak value and valley value of a time-dependent quantity.

Pedersen Potentiometer : Refers to a quadrature-type a.c. potentiometer in which two sets of slide-wires and resistance coils have been fed in quadrature through

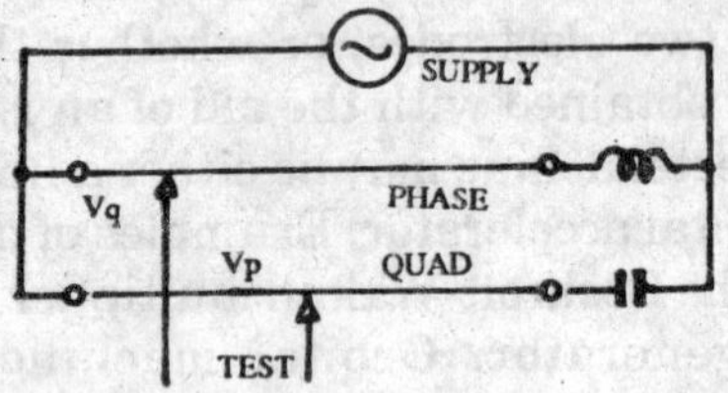

Fig. 3. Principle of Pedersen a.c. Potentiometer.

inductive and capacitive reactances respectively to secure a 90° displacement (Fig. 3).

Peltier Effect : Refers to a thermoelectric effect. When an electric current has been maintained across the junction of two dissimilar metals, the junction temperature alters unless heat is supplied or removed by external means. The rate of which heat must be supplied to the junction to keep its temperature constant has been proportional to the current, and changes sign when the current direction gets reversed. If the current flowing across a junction brings about an absorption of πi watts, π is called the Peltier coefficient.

Pelton Wheel : An impulse-type water turbine which is developed from the simple water wheel. It is having a series of buckets which are placed round the perimeter of a disc, two roughly hemispherical buckets being used at each position with a cutaway portion in the centre to disallow obstruction to the jet of water by the succeeding pair of buckets.

Pentode : An electron tube or valve which is having five electrodes.

Pentode Ignitron : Refers to an ignitron for high-voltage applications. It is having three grids in addition to its a node and cathod. It is used for high-voltage rectification and inversion, or as a high-voltage switch for capacitor, switching in special pulse applications.

Permalloy : An alloy of nickel and iron over 78% of nickel. It is having a high permeability and low coercivity, and hence low hysteresis loss.

Permeability : Short term used for absolute permeability.

Permeameter : An instrument which is used for measurement of the magnetic characteristics of ferromagnetic materials.

Permeance : Refers to the magnetic flux per ampere-turn of total magnetomotive force in the path of magnetic field. It has been the reciprocal of reluctance.

Permeance Coefficient : Refers to the ratio of area of length of a non-magnetic path. If permeance is A, permeance coefficient λ, and absolute permeability of free space μ_0, then we have

$$A=\mu_0\lambda$$

Permittivity : Short term used for absolute permittivity.

Persistence : Refers to the gradual decay of luminescence of a material after the removal or reduction of the causing stimulus.

Perspex : Trade name for a polymethyl methacrylate resin. The outstanding characteristics of Perspex have been its stability, relatively, high softening point and clarity.

P.F. : See power factor.

Phase Advancer : A machine which is able to provide an e.m.f. for injection into the rotor circuit of a slip-ring induction motor to give magnetising volt-amperes, so that the power factor at the stator terminals gets improved.

Phase Angle : Refers to the angles a in the sinusoidal quantity.

$$a=A\sqrt{2}\cos(\omega t+\alpha)$$

Phase-change Coefficient : Refers to the change of phase angle of a sine signal which gets passed through a quadripole; or refers to the change per unit length of an electromagnetic wave in its progress along a transmission line, through a material medium, or through free space.

Phase Changer, Phase Converter : A machine which is used for converting alternating current of a certain number of phases to alternating current of a different number.

Phase Difference : Refers to the angle by which a sinusoidally varying quantity gets displaced in time from another pf of the same frequency.

Phase-equaliser : Alternative name for current-sharing inductor.

Phase Modifier : A synchronous machine which gets arranged for control of its excitation to take leading or lagging reactive volt-amperes so that a load may be having its effective power factor modified.

Phase Modulation : Refers to the imposition of a signal on to an alternating carrier wave in such a wave that the instantaneous phase of the carrier gets varied by an amount proportional to the amplitude of the modulating signal.

Phase-rotation Indicator : Refers to an instrument which is used to indicate phase rotation comprising an iron base with three short iron poles each carrying a coil connection to one lead, having a common neutral point which is not brought out. Pivoted above the magnetic pole has been a light iron disk which will revolve when the coils get connected to a three-phase supply. The direction of rotation will depend on, and indicate, the phase sequence of the supply.

Phase-sequence Components : Refers to the components into which the current values get resolved in analysis of an unbalanced system by method of symmetrical components.

Phase-shift Distortion : Refers to a change in waveform, where the input components are altered in their relative phase due to a non-linear phase characteristic which does not give rise to a phase shift that has been an integral multiple of 2π radians. This is also termed at delay.

Phase Voltage : See star voltage.

Phasor : A term which is used to describe a complex quantity which denotes the combination of r.m.s. value or amplitude of a sinusoidal quantity and its phase angle.

For example

$$A = A \exp (j\alpha)$$

has been the (r.m.s.) phasor of the sinusoidal quantity

$$a = A\sqrt{2} \cos (\omega t + \alpha)$$

Phase-shifting Transformer : A transformer which is used where it becomes necessary to alter phase. Two types have been developed (see Fig. 4). The first has been based on a 3/12 or 3/24-phase transformer. This arrangement provides voltage spaced every 30° for a 3/12-phase or every 15° for a 3/24-phase transformer.

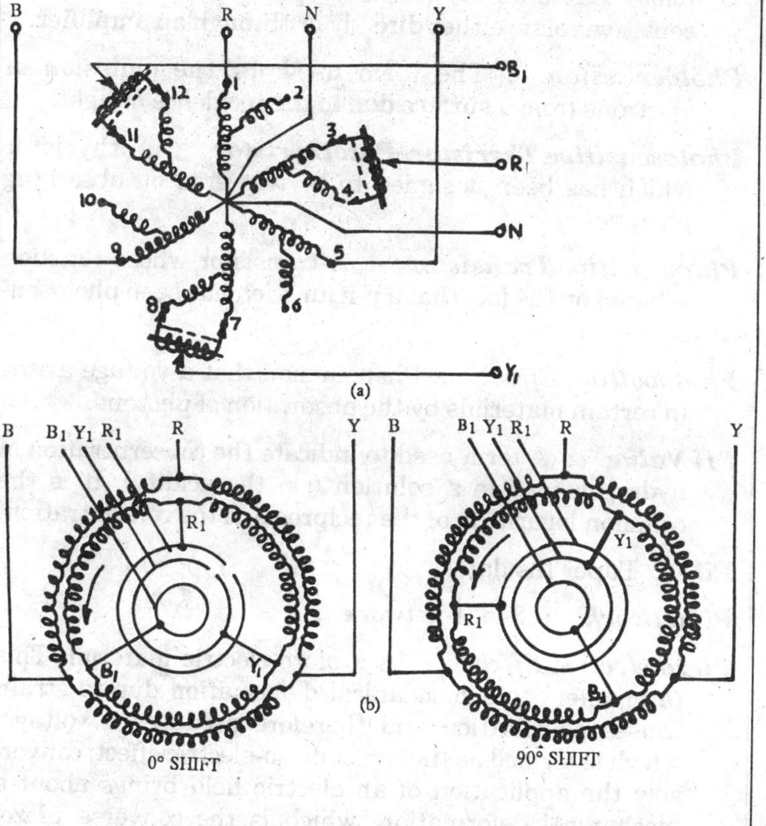

(a)

(b)

0° SHIFT

90° SHIFT

Fig. 4. Phase-shifting transformer :
(a) 3/12-phase-shifter;
(b) induction-motor type phase-shifter.

Phenolic Resign : Thermosetting resin which is used in varnish form to impregnate the filler materials in the majority of thermosetting laminates employed in the electrical industry. It plays an important part in insulators for electric generation and distribution, particularly in switch and control gear.

PH Meter : It is a d.c. valve voltmeter which is arranged to measure the electromotive force produced in a half cell, the electrolyte of which has been the solution on test, and hence provide indication of the pH value of the solution.

Photoelectric Relay : Refers to a combination of a photoelectric device which is operating a conventional sensitive relay, either directly or through an amplifier.

Photoemission : The term used for the emission of electrons from a surface due to the incidence of light.

Photosensitive Thyristor, Phtothyristor : A thyristor which has been designed to be triggered on absorbing photons.

Phtosensitive Transistor : A transistor whose function is based on the fact that a p-n junction has been photosensitive.

Photovoltaic Effect : Phenomenon that a voltage arises in certain materials by the absorption of photons.

PH Value : A term used to indicate the concerntration of hydrogen ions in a solution (*i.e.* the acidity). It is the common logarithm of the reciprocal of the concentration.

Pi. : Paper Insulator.

Pi-network : See II-network.

Piezo-electric Effect : In a piezo-electric material. This term refers to a mechanical deformation due to strain causes polarisation and therefore produces a voltage, which is termed as the direct piezo-electric effect; conversely the application of an electric field brings about a mechanical deformation, which is the converse piezo-electric effect.

Piezo-electric Material : A material that undergoes the piezo-electric effect. it may be a pieao-electric crystal (either of the mineral type like quartz or of the artificially grown type like Rochelle salt, ammonium dihydrogen phosphate and lithium sulphate) or it may be a piezo-electric ceramic, which is having a large number of small crystallites sintered together at a high temperature.

Pilot Controller : Refers to a multi-way switch which is used for controlling the operation of a set of contractors. It is sometimes termed as a master controller.

Pilot Wire : Refers to a conductor in a power system which is used for auxiliary purposes like measurements, communications or protection.

Pilot-wire Protection : Refers to a fact-acting form of network protection. When used in conjunction with over-current protection it will be able to reduce operating times and alleviates difficulties of grading.

Pin Insulator : Refers to an insulator in an overhead-line system which is supported from the corss-arm by a pin on which it is rigidly mounted. It has been usually below the phase-conductor that it supports. It has been suitable up to 33 kV, and has been used for 66 kV (see Fig.5).

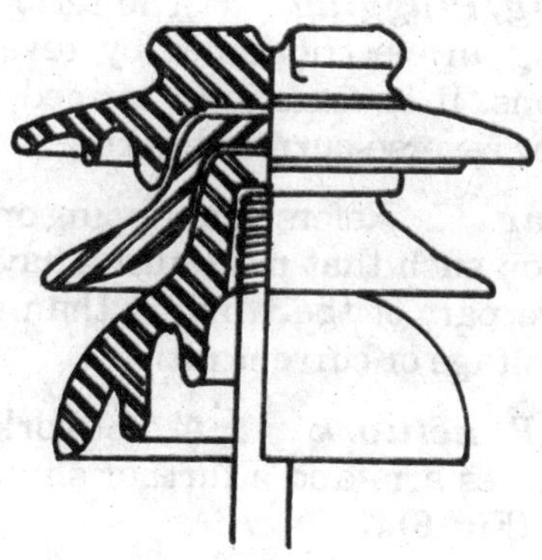

Fig. 5. Pin Insulator

Pin Winding : Refers to a hand-applied concentric winding which becomes necessary in certain types of machine.

Pinch : In a lamp, this terms refers to the airtight seal round the wires connecting the cap contacts to the filament or elecrode.

Pinch Effect : Refers to the mechanical force that tends to urge the current in a wire to flow along the axis of the wire.

Plain-break Circuit-breaker : An oil circuit-breaker does not have arc-control device.

Plasma : Refers to a gaseous mixture of atoms or molecules, or both, with free electrons and ions which arises either in gas conduction or in nuclear fusion.

Plastic : Refers to organic substances having useful combinations of electric and mechancial properties. They are plastic under pressure and heat, and can be moulded, extruded, etc. They are in constant use in the electrical industry.

Plate Electrode : Refers to an earth electrode which is generally of cast iron, mild steel, or copper, not larger than 1 to 2 m square.

Plate Frame : Of a nickel-iron accumulator, this term refers to the nickel-plated steel framework that is able to support the perforated steel tubes containing the active materials of the electrode.

Plug : A connector which has been designed to make contact by insertion in a socket.

Plug Braking, Plugging : The term used for a method of braking an electric motor by reversal of the supply connections. It is sometimes termed as counter-current braking or reverse-current braking.

Plug Tapping : Refers to a tapping on a winding, having its position such that more turns have been included in the active part of the winding than are needed for the service voltage or current ratio.

Π-network, Pi-network : A network having a shunt arm, a series arm and a further shunt arm equal to the first (see (Fig. 6).

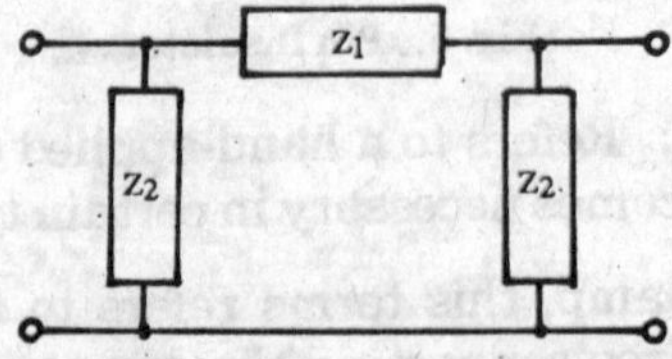

Fig. 6. A symmetrical unbalanced Π-network.

P-n Junction : The term used for the junction between a *p*-type semi-conductor and an *n*-type semiconductor which is having rectifying properties. If a voltage is applied as shown in Fig. 7(*a*), electrons in the *n* side will get attracted across the junction to the positively biased terminal and holes in the *p* side to the negatively biased terminal, and a large current will start flowing. Now if the polarity gets reversed, (*b*) both electrons and holes will get attracted away from the junction until the electric field produced by their displacement would be equal to the equal applied field. Only a small current would flow, due to free electrons and holes being produced near the junction by thermal agitation.

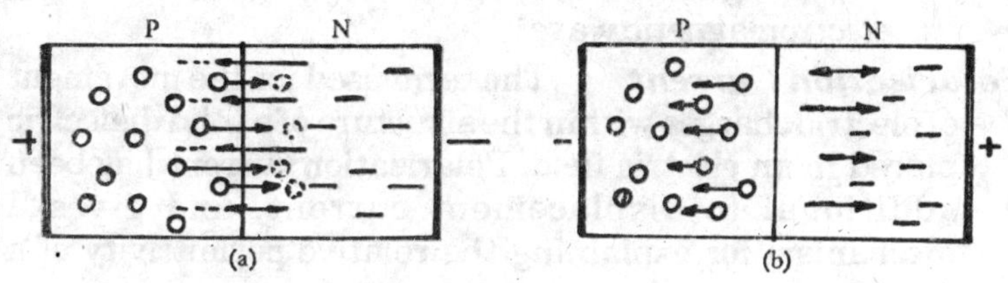

Fig. 7. Rectifying action of a p-n junction : (a) forward bias; (b) inverse bias.

P-n-p Transistor : A transistor which consists basically of two *p*-type areas with a thin *n*-type layer between them (Fig. 8).

The first *p* region has been the emitter, the *n* region the base, and the second *p* region the collector. There are thus two *p-n* rectifying junctions.

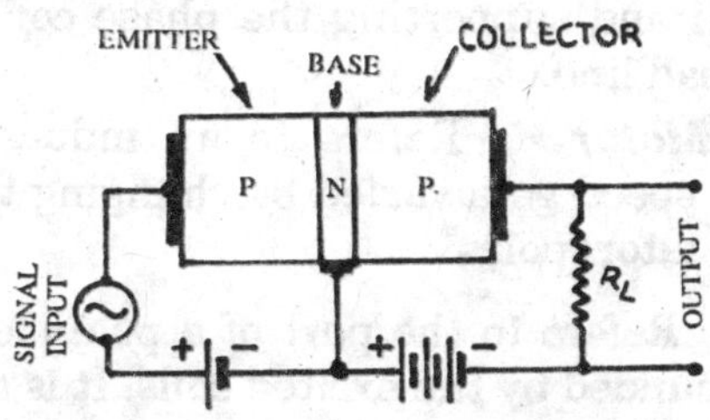

Fig. 8. Basic p-n-d transistor.

Point-contact Transistor : A transistor in which the base has been a crystal of p-type or n-type semiconductor and the emitter and collector have been in effect, small areas of material of opposite type which are formed round the points of two closely spaced electrodes. It has been now largely superseded by the junction transistor.

Polarisation :

1. Means the orientation of the positive and negative mole-cular charges in a dielectric material situated in an electric field.
2. Also, refers to the formation of a film of gas or a strong ionic concentration at an electrode of a primary cell, causing a reduction in the electromotive force.
3. Also, refers to the direction of the planes of the electric and magnetic components of the field of an electromagnetic wave.

Polarisation Current : The term used for the movement of electric charges within the structure of a solid dielectric placed in an electric field. Polarisation current has been additional to displacement current, and gives a mechanism for explaining the relative permittivity of a dielectric material.

Pole :

1. Refers to each of the lines or terminals of a circuit or equipment between which a large voltage exists.
2. Refers to the part of the magnetic circuit of a machine between the yoke and the air-gap (otherwise known as the pole piece).
3. Refers to the terminal or accessible part of an electrode in a voltaic cell.
4. Refers to an upright beam which is inserted in the ground and supporting the phase conductors of an overhead line.

Pole-change Motor : Refers to an induction motor in which the speed gets varied by changing the number of primary (stator) poles.

Pole-Core : Refers to the part of a pole piece which has been surrounded by the excited coils. It is also known as the pole shank.

Pole Horn : Refers to the part of a pole piece that projects circum-ferentially beyond the excited coils.

Pole Modulation : Refers to a method of winding a pole-change motor which provides a speed ratio within the range 1.5 : 1.

Pole-mounted Transformer : Refers to a transformer substation which is mounted directely on a pole of the overhead transmission line. Transformers up to 100 kV A three-phase could be counted on single poles and up to 200 kV A on H-poles.

Pole Piece : Refers to the part of the magnetic circuit of a machine between the yoke and the air-gap. It is alternatively termed as the pole or magnet pole.

Pole Shoe : Refers to the separable part of a pole piece that faces the armature of a machine.

Pollack Construction : Refers to a method of fitting long commutator bars for eliminating any lifting due to centrifugal force.

Polyester Resin : It is a synethetic thermosetting resin which may be used in encapsulation. It has been also used with various fillers and in the production of reinforced plastics. Polyester/glass-fibre fabrication have been used for cable ducts.

Polyethylene : A thermoplastic polymer of ethylene, also known as polythene. Two major advantages have been its extremely low permittivity and power factor. Its break-down voltage has been high, and its volume resistivity considerably higher than those of other important insulating materials. The properties of polyethlene have made it an ideal material for wire-and cable-covering.

Polymethyl Methacrylate : Refers to a rigid glass-clear transparent resin which is of good optical, mechanical and electrical properties.

Polyphase System : Refers to an a.c. network to which have been supplied two or more electromotive forces of the same frequency but displaced in time phase. Asymmetrical M-phase system of e.m.f. is having M e.m.f.s which are equal in magnitude, wave-form and frequency, and separated in time phase by 1/M of a period. Such e.m.f.s. are normally produced in an electro-magnetic machine having M identical phase windings displaced from each other by 2/M of a pole-pitch. The voltages are then as follows:

$$V_A = V_m \sin \omega t$$

$$V_B = V_m \sin (\omega t - 2\pi/M)$$

$$V_C = V_m \sin (\omega t - 4\pi/M)$$

....

$$V_M \triangleq V_m \sin \{\omega t - (M-1)2\pi/M\}$$

where V_m denotes the peak voltage of each phase A,B,C......M. The sum of the voltages at every instant has been zero:

$$V_A + V_B + V_C + ... + V_M = 0$$

Polystyrene : Refers to a thermoplastic material whose electrical properties have made it suitable for moulded high-frequency insulating components. Insulating film has been produced from it, and as a moulding material is has been used for coil formers, transformer bobbins, etc.

Polytetrafluoroethylene : A thermoplastic material which finds increasing use in the electrical industry.

Polyvinyl Chloride : A thermoplastic material which has low water absorption and is resistant to corrosion and mechanical abrasion. It finds use in the electrical and associated industries for extruded cable insulation and sheathing.

Pool Cathode : A liquid cathode usually mercury, which is used in a mercury are rectifier.

Porcelain : Insulative material which is produced from china clay with quartz and feldspar. It finds use in the manufacture of high-voltage insulators.

Port : In any network, this term refers to the combination of two terminals such that the current passing into the network through one of the two has been identical with the current leaving the network through the other. Terminal pair has been an alternative name.

Porter-bentley Discriminator : A crane load discrimina- tor using a series diverter rheostat which mechanically gets operated by a deflection proportional to the torque being exerted.

Postive Booster : A booster which is arranged to increase the voltage supplied from another source.

Positive Feedback : Means the return of energy from the output of an equipment to the input, in phase with the input or of the same polarity. It could be used for increasing the gain of an amplifier or for generating oscillations. It is also termed as regenerative feedback.

Positive Phase Sequence : Refers to that phase sequence which is corresponding to the direction of rotation of a polyphase machine.

Positron : A 'positve electron', *i.e.* an elementary particle which has a mass similar to that of the electron but has been carrying an equivalent positive charge.

Post Insulator : An insulator which has been built in the form of a post. It has been used for supporting busbars in an h.v. outdoor substation, and has been found to be suitable for use in heavy-industrial districts with contaminated atmospheres.

Post Office Bridge : Refers to a self-contained resistance bridge which was developed originally by the Post Office. It has been a form of Wheatstone bridge.

Potential : At a point, this term refers to the potential difference between that point and earth, the latter being regarded at zero potential. The unit of potential has been the same as that of potential difference.

Potential Difference : Refers to the line integral of the field strength along a freely chosen path between two points in an electric field that gets produced in the absence of any electromagnetic induction. Its SI units is the volt.

Potential Divider : See voltage divider.

Potential Gradient : Refers to the potential difference per unit length in the direction in which it has been a maximum. It is measured in volts per unit length.

Potential Transformer : See voltage transformer.

Potentiometer : A part of an instrument which is used for comparing voltages, needing in addition a cell or battery to supply a steady direct current, a standard cell and a null-detection galvanometer; it would then measure an unknown direct electromotive force or potential difference within its range, there being no current taken from the test source at balance. Basically it consists of a length of uniform resistance wire which is stretched against a scale calibrated in length units. The position of a sliding contact has been read against the scale. The elementary circuit is depicted in Fig. 9, where a low voltage supply feeds the slide-wire.

Potentiometer Control : Refers to the form of electric braking frequently used on d.c. motors in crane control. The series motor is connected generally as in Fig. 10.

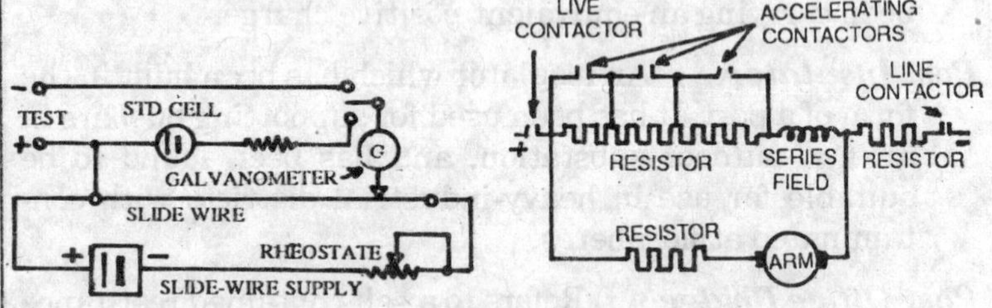

Fig. 9. Principle of an elementary d.c. potentiometer.

Fig. 10. Typical main connecdtions giving potentiometer control.

Potentiometer-type Field Rheostat : Refers to a field regulator in which the resistor may get connected across the source of supply, means being provided for the field winding to et connected between various points on the resistor.

Potier Reactance : Refers to the reactance of a synchronous machine as determined by a specific graphical construction which gives also the armature reaction and regulation from the open-circuit, short- circuit and zero-power-factor characteristics.

Power : Refers to the rate at which work has been done or energy gets converted from one form to another. The basic unit of energy has been the joule, and the joule per second, or watt, has been the basic unit of power. The watt, with its multiples and sub-multiples, forms the main column of the diagram.

Power Angle : Alternative name for load angle.

Power Chart : Refers to a chart for a synchronous generator, motor or condenser from which its performance in relation to stability, permissible loading and excitation could be readily deduced.

Power Component : Alternative name for active component.

Power Factor : Refers to the ratio of average power dissipation to the apparent power (*i.e.*, the product of r.m.s. voltage and current) in an a.c. network or part there of. If both voltage and current have been sine functions and under steady-state conditions are having a phase-displacement ϕ, then the power is $P=VI\cos\phi$, the volt-ampere product has been VI, and the power factor has been as follows :

$$P/VI = \cos\phi$$

Power-factor Correlation :. Use of equipment to restore a lagging power factor to near unity. Synchronous compensators or static capacitors may be used or a compensated motor used.

Power-Factor Meter : Refers to an instrument which is able to measure the difference in phase between two electric quantities of the same frequency, normally the voltage and the current in a circuit. It is sometimes termed as a phase meter.

Power Selsyn : Refers to a selsyn which has been designed to ensure synchronous rotation of two or more mecha- nisms, the relative angular displacement varying with the transmitted torque.

Power Signalling : The operation of railway points and signals by electric or electro-pneumatic power.

Power Station : Refers to an installation, which includes its housing, an assemblage of various apparatus with their accessories. It is designed to produce electric energy in substantial quantities from an naturally occurring form of energy.

Power System : A technical system which is used for the supply of electricity. This system includes one or more power stations and systems for the transmission and distribution of electric energy.

Power-system Capacitor : A capacitor which included in a power system so as to improve the characteristics of that system. Capacitors have negative reactance and may get connected in series with the phase of an a.c. circuit ot compensate the inductive reactance of the system. Capacitors so used are called series capacitors. When capacitors are connected across the phases parallel to an inductive load, they draw leading reactive power which compensates the lagging reactive power drawn by the load, thereby improving the overall power factor. Capacitors so connected are called shunt capacitors.

Precipitator : See electrostatic precipitator.

Precision Instrument : An instrument which has been designed for measurements requiring high accuracy.

Prece's Law : A law which relates the diameter of a wire to its fusing current and temperature. If the diameter has been d, current I and melting point T, then $T = aI^2/d^3$ or $I=kd^{3/2}$. Thus if we know k for a wire of given diameter, the fusing current may be calculated.

Prefocus Lamp : The term used for a lamp in which the distance between the filament and the lampcap has to set very accurately during manufacture.

Pre-impregnated Cable : Refers to a paper-insulated cable, the paper tapes of which get impregnated before assembly.

Prepayment Meter : It is domestic form of energy meter which incorporates a coin mechanism, whereby the

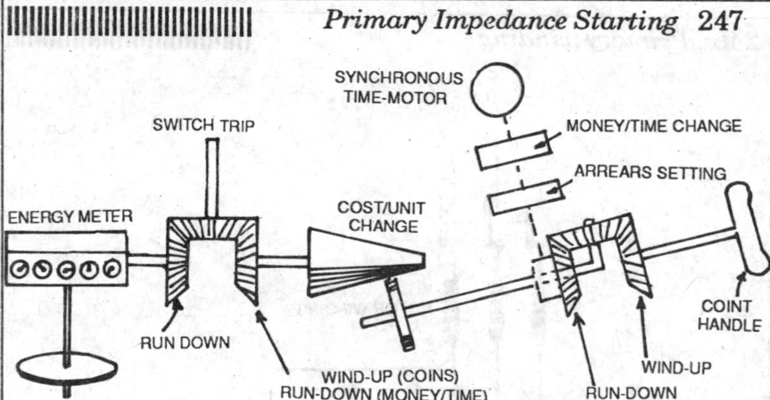

Fig. 11. Mechanism of a prepayment meter.

provision of energy has been contingent upon the prior insertion of a suitable coin. The prime essential of this meter has been the counting of energy units (kWn) against coin value in accordance with a present tariff. Fig. 11 shows the essential mechanism schematically.

Pressel Switch : A switch which has been designed for attachment to a fiexible cord. It is alternatively known as a pendant switch, pear switch or suspension switch.

Pressurised-water Reactor : Refers to a type of nuclear reactor that employes natural water as the coolant (and possibly the moderator also) at temperatures of 200-300°C.

Preventive Reactor, Resistor : Terms now replaced by transition inductor, resistor.

Price's Guard Wire : In insulation testing, it is a conductor which is placed round the edge of the insulating material to disallow the flow of earth leakage current through the measuring instrument.

Primary Cell : Refers to an electrolytic cell, in which two electrodes of different conductive materials associated with an electrolyte produce an electrochemical e.m.f.

Primary Distribution : Refers to high-voltage distribution from generating stations or main substations to load centres or secondary sub-stations.

Primary Impedance Starting : Refers to a method of reducing the starting current of a motor, using an impedance in sereis with the primary winding of the motor.

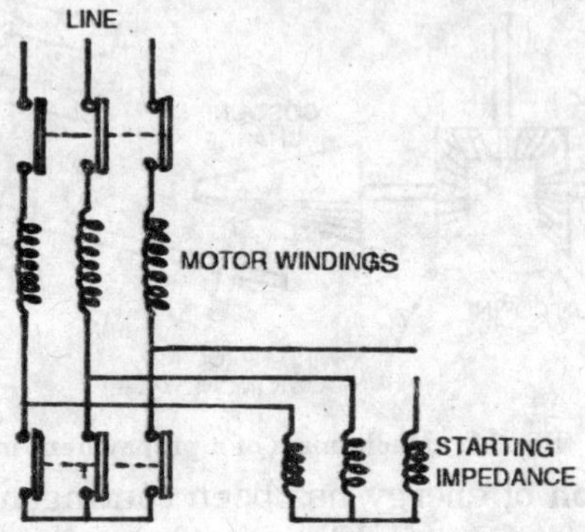

Fig. 12. Elementary diagram of connections for primary impedance starting of a large high-voltage squirrel-cage motor.

The impedance will generally be a resistance but may be a reactance if it has been desired to keep losses in the control cubicle to a minimum. This starting scheme provides the lowest torque per ampere of any, but gives smooth starting with a single step; as the motor accelerates and its current falls, the voltage drop in the series impedance falls also, so that more voltage has to be applied to the winding (Fig. 12)

Primary winding : Refers to that winding of a transformer which has been connected to the source of energy.

Printed Circuit : Refers to a circuit, including its connections and also certain of its components which is produced by an etching process on metal foil deposited on an insulting sheet material as base.

Projection Welding : Refers to a development of spot, welding, in which the current has been concentrated at the desired points by projections on one of the components. Flat electrodes have been used and a number of welds could be made simultaneously.

Proofed Tap : Refers to a cotton-cloth tape coated with a rubber compound which gets wrapped round the insulation of rubber insulated cables.

Propagation Coefficient : Refers to a combination of the attenuation coefficient, α, and the phase-change coefficient, β, in the form $y = \alpha + j\beta$.

Propeller Turbine : Refers to a water turbine which is used for low heads of water. It is a development of the reaction or Francis turbine.

Proportional Counter : Ionisation counter which operates at a voltage sufficiently low for the pulse of current to be small and proportional to the incident radiation.

Prospective Current : Refers to the current that would flow on the making of a circuit when the circuit has been equipped for the insertion of the fuse but the fuse has been replaced by a link of negligible impedance.

Protective Box : A box which is used to enclose lead joint boxes and sleeves, and afford mechanical protection to a joint. It could be of cast iron, earthenware or concrete and has been usually filled with bitumen.

Protective Device : A device which has been designed to disallow any damaging effect on an electric device, equipment or system when there arises an abnormal condition.

Protective Gap : Refert to a form of surge diverter which consists of two electrodes separated by an air space. It has to be connected between the conductor and earth to divert ot earth part of any surge which may take place.

Protective Multiple Earthing : Refers to the method of earthing which is used in low-voltage distribution using the neutral for the return path to the transformer. The earthing of the neutral conductor is itself done at intervals along the line, usually at consumer's premises and at the end of each distribution.

Protector Tube : Alternative name for expulsion gap.

Proton Resonance : Refers to resonance which is associated with the orientiation of proton spins. Its frequency is employed to measure the strength of a magnetic field to a very great accuracy.

Proton Synchrotron : See synchrotron.

Proximity Effect : Refers to the effect of one current-carrying conductor on another. If two or more neighbouring conductors carry currents (as in a coil), the current distribution in any one conductor gets affected by the magnetic field produced by the adjacent conductors. The current density becomes greatest in those parts of a conducor which are encircled by the smallest flux.

Proximity Switch : A switch which is operated without any physical contact. It incorporates a sealed head having a sensor unit which is able to detch the presence of magnetic materials and sends an impulse to operates a relay, etc.

P.I.F.E. : Polytetra fluoroethylene.

p-type Semiconductor : Refers to a quadrivalent semiconductor which is including a suitbale trivalent acceptor element so that the crystal lattice is having a number of holes, or positive charge carriers.

P.u. : per unit system

Pulsactor : Refers to a saturating reactor used with a capacitor and transformer to supply positive and negative pulses for the control of an ignitron.

Pulsating Current : Refers to a unidirectional current, the magnitude of which gets varied rhythmically with time.

Pulse : A stimulus or signal of short duration ; it is generally of rectangular form.

Pulse Modulation : Refers to modulation in which the carrier has been a series of pulses, some characteristic of which gets varried in accordance with the modulating signal. There are three types of pulse modulation

(a) *Pulse amplitude modulation:* : When the modulation signal gets sampled at regular intervals; the amplitude would be proportional to that of the modulating signal.

(b) *Pulse-width modulation:* : When the amplitude and mean repetition frequency have been maintained constant by the width of the pulses get varied in accordance with the amplitude of the modulating signal.

(c) *Pulse-position modulation:* : In this the amplitude, width and mean repetition fequency have been maintained constant but the position of the pulse has been varied in accordance with the modulation. In this case timing pulses are required so that the variaitons in position may be determined.

Pulse Reflection Test : A method which is used for localising a fault in a cable, of particular application to telecommunication circuits.

Pulse-type Regulator : Refers to an automatic voltage regulator in which the excitation current has been fed into the exciter field in a series of pulses, which can vary in magnitude and/or frequency. It is usually employed for medium-sized machines of capacity up to about 6000 kV.

Pulverised-fuel Boiler : A boiler used with generating plant in which coal gets reduced to a fine powder and projected into the combustion chamber by using a current of hot (primary) air. Further preheated (secondary) air to make up the required amount of combustion is also blown in separately. The ensuing turbulence in a high-temperature chamber is able to facilitate through combustion.

Pump Line : In a railway train, this term refers to a cable (or train line) which extends throughout the length of the train, for the control of auxiliary apparatus such as air-compressors.

Pumped Storage : Refers to a method of increasing a power system load during periods of low electrical demand, by pumping water into storage reservoirs. This water may later be employed for generation during periods of peak load.

Punching : Of a machine or transformer this term refers to each of the thin, metallic sheets which form part of the core. The sheets are also termed as core plates, laminations, or stampings.

Punch-through : Refers to a form of transistor breakdown which takes place when the depletion layer of the collector barrier gets moved through the base region to make contact with the emitter.

Puncture : The term used for a disruptive breakdown through a solid insulating material.

Puncture-withstand Test : A test which is used to test equipment designed to operate at high voltages. This test object has been immersed in insulating oil and subjected to a specified voltage, for the minimum time necessary to measure the voltage. This test has to be withstood flashover or puncture.

P.v. : Peak-to-valley value.

P.v.c : Polyvinyl chloride.

Pyrochlor : A fire resistant flued which is used as a filling medium for transformers. As it is also non-flammable, it will not produce explosive gases or mixtures with air when heated or exposed to an electric arc. It could be operated safely at high temperatures for sustained period.

Pyrometer : An instrument which is used for the electric measurement of temperature.

Q-factor : The term used for a figure of merit for a resonant device like an LC circuit, a cavity resonator, a piezo-electric crystal or, by analogy, any related device in the mechanical field. Such system get resonated by cyclic interchange of stored energy, accompanied by energy dissipation due, for example, to resistance, electromagnetic radiation or friction. If ω denotes the angular frequence of resonance, then

$$Q = 2\pi \frac{\textit{energy stored}}{\textit{energy loss per cycle}}$$

$$= \omega \frac{\textit{energy stored}}{\textit{average power dissipation}}$$

For an inductive circuit, $Q = \omega L/R$; for a capacitive circuit. $Q = 1/\omega CR$.Q factor has been the abbreviation for quality factor.

Quad Cable : A cable having four separately insulated conductor twisted together.

Quadrant Electrometer : A basic form of electrometer.

Quadrature Axis : In an electromagnetic machine, this term refers to the axis electrically at right-angles to the direct axis. It is also called the cross axis.

Quadrature Component : Alternative name for reactive component.

Quadripole : This term has been replaced by two-port network, with the alternative term two-terminal network.

Quantity of Charge : Refers to the product of a constant direct current in a circuit and the time during which it gets maintained. The SI unit has been the coulomb (symbol C). The ampere-hour (A-h) has been a unit frequently used.

Quarter-phase System : Refers to a two phase system. It is also called because the displacement of one phase with respect to the other has been one quarter of a period.

Quartz : Refers to a piezo electric crystal. Plates cut from quartz are generally used as circuit elements in oscillators designed to provide the stable frequency needed in most radio, transmitters.

Quench Voltage : Refers to the voltage at which gas conduction ceases when the applied voltage between two electrodes gets reduced.

Quick Make-and-break Switch : Refers to a switch that operates rapidly to make or break a circuit, the rate of motion of the blade contacts at the critical moment of contract or separation being independent of act.on of the operator. It is also called a snap switch.

Quickstart Tube : It is a fluorescent tube, the axterior of which has been coated by a silicone treatment.

Quill Drive : Refers to a form of drive which is used with electric traction. It consists of a hollow shaft surrounding the driving axle and having sufficient clearance to allow the necessary relative movement between the spring-borne quill and the axle. The quill is having a gearwheel engaging with a pinton on the motor shaft. A twin or double armature motor is used and the secure flexibility the pinions may be spring cushioned.

R. : Symbol for Rontgen.

Rad : (1) Unit of absorbed dose of radiation; 1 rad=10^{-2} J/kg.(2) symbol for radian, which is the SI unit of plane angle.

Radar : A technique which uses radio waves for the detection of objects and the determination of their direction and distance from a reference point. The process is using waves of very short wavelengths, propagated as a narrow beam by using a highly directional aerial and reflected off the objects.

Radial Feeder : See independent feeder.

Radial System : Refers to a distribution system in which single supplies have been given as individual lines from a central substation, or as spurs from a main distributor.

Radiant-arc Furnace : Refers to a modification of the Heroult arc furnace. The electrodes have been so arranged that the arc gets deflected downwards away from the roof on the furnace charge (Fig. 1)

Radiation pyrometer : Refers to a pyrometer in which radiation energy of all wavelengths gets forcused by using a quartz lens or concave mirror on to the hot junctions of a small, compact thermopile, the e.m.f. being measured potentiometrically or by using an indicating instrument calibrated to read temperature direct.

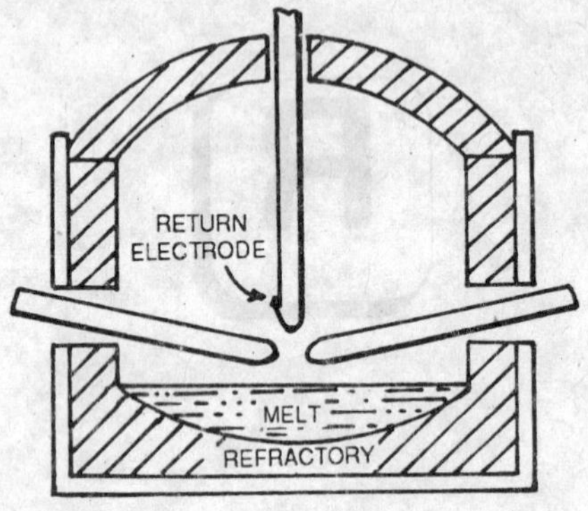

Fig. 1. Principle of the radiant-arc steel furnace.

Radio-frequency Heating : See high-frequency heating, dielectric heating, induction heating.

Radio Interference : The term used for the reduction of the quality of receiption of a desired radio signal by undesired electromagnetic radiation originating in natural or man-made phenomena. Equipment for interference suppression has been used at the source where repetitive sparking takes place, for example in electric machines having commutator or in combustion engines with spearking plugs.

Radio Waves : Electromagnetic radiation in the approximate frequency range 10^3 to 10^{13} Hz.

Radiograph : Refers to an image which gets formed on a sensitive film or plate by the action of X-rays of gamma rays which have passed through an object. It is sometimes called a skiagraph or rontgenogram.

Radiotherapy : Refers to the electromedical treatment of diseases, using radiation.

Ramp Function : Refers to a function of time, which grows uniformly from zero at t=O and given at any instant t thereafter by

$$f(t)=kt$$

The ramp function finds use in transient analysis, as it can represent the effect of a steep linear wavefrom, or the initial part of a slowly rising sine function.

Random Winding : Alternative name for much winding.

Rate of Rise of Restriking Voltage : Refers to the rate at which the voltage across the contacts of an interrupting device rises.

Rated Value : Refers to a value which is assigned by the manufacturer for a specified operating condition of a device or equipment.

Rating : Refers to the whole of the rates values and operating conditions of a device of equipment.

Radio Adjuster : Alternative name for a tap changer.

Ratiometer : Refers to an instrument, the operation of which is dependent on the ratio between the currents in two separate circuits or branches of one circuit.

Reactance : The term refers to the voltage/current ratio in r.m.s. terms of a pure energy-storage circuit parameter (inductance L or capacitance C or any combination thereof) under strictly sinusoidal conditions and at a given angular frequency $\omega=2\pi f$. For a pure inductor the reactance has been

$$X_L=V/I=\omega L=2\pi f L$$

For a capacitor it has been

$$X_c=V/I=1/\omega C=1/2\pi f c$$

In each case V and I denote respectively the r.m.s. voltage and current.

Reactance Voltage : In a circuit having reactance, this term refers to the product of the current and reactance.

Reaction a.c. Generator : An alternator which has salient poles without field windings. The exciting current is got from an independent a.c. source at the required frequency.

Reaction Turbine : Refers to a steam turbine in which the blading has been modified to give pressure compounding with the steam expanding in both fixed and moving blades so that its velocity also increases as it gets passed through the latter.

Reactive Component : Refers to the component of an alternating current or voltage (considered as phasor quantities) which has been in quadrature with the current or voltage.

Reactive Factor : Refer to the ratio of reactive power to apparent power.

Reactive Load : Refer to a load in which the sinusoidal current and voltage have been out of phase at the terminals.

Reactive Power : Refers to the product of the current and reactive component of the voltage, or of the voltage and reactive component of the current, in each case under sine conditions (symbol : O), Unit of reactive power has been the reactive volt-ampere (symbol : V Ar).

Reciprocal Ohm. : Obsolute unit, now the SI unit siemens.

Reciprocity Theorem : A theorem whichis used in netweork analysis. If a given electromotive force in branch A of a network produces a current in which branch B, then the same e.m.f. introduced into branch B would produced the same current in branch A. The ratio of the e.m.f. and the current in such a case is termed as the transfer impedance.

Recording Instrument : Measuring instrument which represents the reading as a permanent record, such as on a chart showing the value at any time.

Recovery Voltage : Refers to the sustained mains-frequency voltage of a power system that appears across the terminals of a circuit-breaker after interruption of the circuit.

Rectifier : A device which is used for converting an alternating current or oscillating current into a unidirectional or approximetaly direct current. This may be attained by the inversion or suppression of alternate half waves.

Rectifier Instrument : Refers to a moving coil instrument which is made suitable for the measurement of alter- nating current by the addition of a metal or semiconductor rectifier.

Recurrent-surge Oscillograph : Refers to a cathode-ray oscillograph whose time-base gets synchronised with a regularly repeated transient voltage so that the trace becomes visible.

Reference Electrode : An electrode which is used in electrochemistry to measure the potential of another single electrode.

Reflecting Galvanometer : See mirror galvanometer.

Refrigerator : A device which is used for obtaining a reduction in temperature. Three ways of achieving this are : (a) Motor-driven compressor. It compresses a gas called the refrigerant. The gas is fed into a condenser where it gets cooled and liquefied, the compressed liquid is led into an evaporator through a pressure-reducing device, such as an expansion value or a capillary tube. In the evaporator the the liquid evaporates and in so doing extracts its latent heat of evaporation from its surroundings. Thus a cooling effect would be produced. (b) Heater-absorber. In this unit, ammonia is made to evaporate in the evaporator by being connected to an absorber which absorbs the ammonia into water and brings about necessary reduction in pressure. The water carrying the ammonia is allowed to pass into a boiler where it gets heated. The heating drives off the ammonia which is then cooled in a condenser and subsequently passed back to the evaporator. (c) Thermoelectric device. It takes advantage of the Peltier effect.

Regenerative Braking : Refers to a method of braking an electric motor using the inherent property of most types of electric motor to generate when run factor than their no-load speed. If the supply has been able to absorb power it constitutes a loss of energy, which causing a braking action. The braking torque tends to increase with speed, providing an ideal braking characteristic for motors that have to deal with overhauling loads.

Regulation : Refers to a particular change of an output quantity with load. For the governor of a prime-mover the regulation has been the change in speed between no-load and full-load prime-mover output, which is normally expressed as a percentage of full load speed.

Regulator Cell : Refers to one of several cells at the end of a battery of accumulators which could be switched into or out of circuit to maintain the total output voltage constant. It is also termed as an end cell.

Reheat : Refers to the improvement of steam-turbine efficiency by extracting the steam after it gets passed through a portion of the turbine and taking it back to the boiler for reheating to a temperature approaching the initial value. It then gets returned to the set to continue the expansion.

Rejector Circuit : Refers to an electrical network which is having a frequency characteristic with an anti-resonance point.

Relative Permeability : For any material, this term refers to the quotient of its absolute permeability by the mag- netic constant.

Relative Permittivity : For any material, this term refers to the quotient of its absolute permittivity by the electric constant.

Relaxation Oscillator : Refers to an electronic oscillator whose frequency has been determined by the charging and discharging of a capacitor through a resistor, or by an analogous process with an inductor. The most important example has been the multivibrator.

Relay : Refers to a device which is so designed that the electric, magnetic, or thermal effect produced when electric power is applied causes a sudden predetermined change in one or more electric circuits.

Reluctance : This term refers to the ratio of magnetomotive force to flux for a magnetic circuit, or part thereof (symbol: R*m*; SI unit 1 per henry).

Reluctane Motor : Refers to a synchronous machine having a wound stator and a case rotor which is constructed to provide two axes of widely differing magnetic reluctance. The resulting two fluxes gives rise to a rotating field which makes the rotor to lock on to the supply frequency.

Reluctivity : Refers to the reciprocal of permeability. It may be a measure of the ability of magnetic material to conduct magnetic flux.

Remanence : See remanent flux density.

Remanent Flux Density : Flux density left in a material after an initial magnetic field strength gets reduced to zero.

Repulsion-induction Motor : Motor having the starting characteristics of a repulsion motor and the running characteristics of an induction motor. The stator has been identical with that of a repulsion motor. The rotor is a drum-type lap winding having commutator similar to the rotor winding of a repulsion motor, but in addition it is having a device for short-circuiting the commutator segments. Alternatively, the motor may have a squirrel-case winding at the bottom lof the rotor slots; this winding takes over at speed and gives induction motor characteristics.

Repulsion Motor : Refers to a single-phase motor in which the field and armature fluxes repel each other to produce a torque in the rotor. If the connections for a single-phase compensated-series motor get changed to those shown in Fig. 2. the compensating winding would act as the primary winding of a transformer and the armature winding as the short-circuited secondary. The magnetomotive forces in the two windings would be nearly equal and opposite to one another. The torque is produced by the interaction between the exciting flux ϕ_e due to the exciting winding E and the armature current I_a, but in

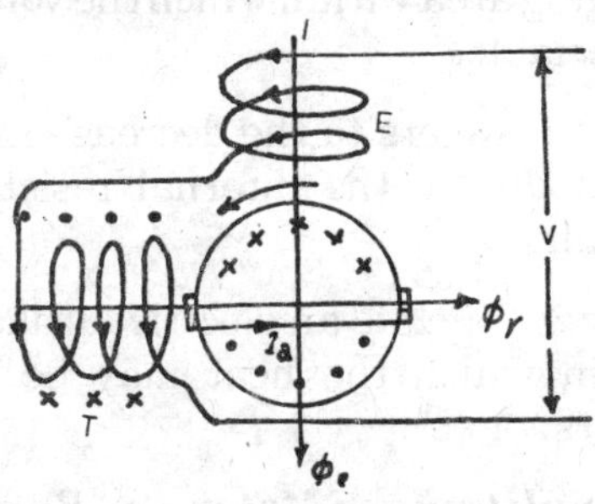

Fig. 3. Repulsion Motor with a single windings.

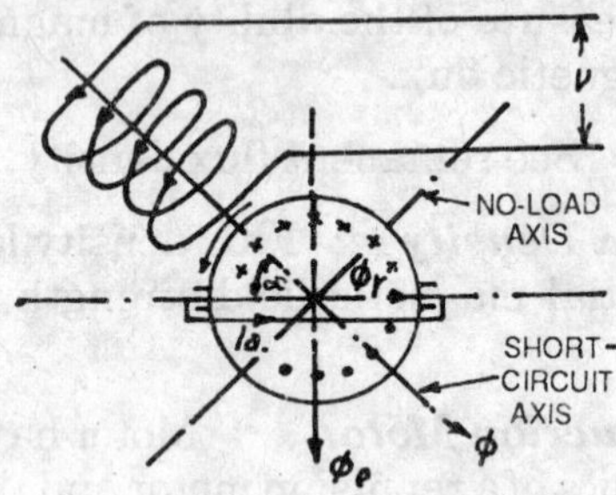

Fig. 3. Repulsion Motor with two winding.

this case the armature current has been not taken directly from the supply but has been produced by induction of transformer action from the winding T, called the transformer winding. The machine with two field winding could be replaced by one having only one winding with its axis at an angle α to the brush axis as in Fig. 3.

Residual Magnetism : Refers to the magnetism which remains in a body after the magnetising force has been removed.

Resistance : Refers to the opposition which is offered by a conducting material (in the widest sense) to the passage of a conduction current. Unit ohm (Ω). It may be defined under steady conditions by the voltage/current ratio of a homogeneous material or any other device capable of conduction.

Resistance Area : Of an earth electrode, this term refers to the surrounding area within which the voltage gradient has been appreciable.

Resistance Drop : Refers to the decrease in output voltage of a circuit due to the internal resistance voltage within the circuit.

Resistance Furnace : It is an electric furnace of the non-melting type in which the heat may be produced by heating resistors.

Resistance-start Split-phase Motor : Refers to the simp- lest and cheapest type of split-phase motor in fractional-horse-power ratings. The rotor has been of the squirrel-case type. The resistance gets incorporated into the auxiliary or starting winding itself, so that no additional equipment, other than the starting switch, becomes necessary.

Resistance-temperature Coefficient : Refers to the rate of change of resistance of a conductor material (or insulation resistance of an insulating material) with temperature. At a temperature $\theta°$, the resistence R_θ compared with that at 0°C is having an expression empirically of the form

$$R_\theta/R_\theta=1+\alpha_0\theta+\beta_0^2$$

Resistance Thermometer : Refers to a thermometer which operates on the principle of increase of resistance with temperature.

Resistance Voltage : Refers to the voltage due to a current which flows through the resistance of a circuit. It has been the product of the current and the resistance.

Resistance Welding : Refers to a process by which metal parts have been welded together, the junction faces being heated by the passage through them of a heavy current followed by their consolidation under high mechanical pressure.

Resistivity : A term which denotes volume resistivity. It was also known as specific resistance.

Resolver : A form of synchro which is having a rotor with two quadrature phases. It is used in radar and for computing.

Resonant Circuit : Refers to an electric circuit that is having capacitance and inductance such that free oscillation may be obtained.

Resonant Frequency : Refers to the frequency of forced oscillation at which resonance takes place in an oscillatory system.

Restriking Voltage : Refers to a high-frequency transient voltage which appears across the contacts of an interrupting device at a current zero. The important characteristics have been the amplitude and the rate of rise of restriking voltage.

Retentivity : Obsolete term. It has now been superseded by remanence.

Return Feeder : Alternative name for negative feeder.

Reverse Compound-wound Motor : Alternative term for differentially compound-wound motor.

Reverse-power Protective System : Refers to an arrange- ment of apparatus respective to a distrubance in a portion of equipment. It is able to solate the appropriate section if the direction of flow of the electrical energy gets reversed.

R.F. : Radio frequency.

Rheostat : Refers to a resistor that has been equipped with means for varying the amount of resistance in circuit.

Rheostatic Braking : Refers to a method of braking an electric motor by negative loading, making the motor to feed power as a generator into a resistance.

Rheostatic Regulator : It is an automatically operated rheostat that controls the main-exciter field current for maintaining constant a.c. generator voltage; today, this type of regulator has been generally used for the control of small and medium-sized generators.

Ring Circuit : Refers to a system of domestic distribution in which, by means of a closed ring circuit, an unlimited number of 13 A socke-outlets could be connected to one 30 A fuse-way serving up to 100 m^2 of floor area.

Ring System : Refers to a distribution system in which the distribution substations get connected in a continuous closed circuit from the main substation.

Ring Winding : A winding which consists of coil wound round an annular magnetic core, each coil being looped through the ring. It may alternatively be called a toroidal winding or a Gramme winding.

Ripple Control : A method of control, *e.g.*, of street lighting, from a central point, with a control system using the supply network as the signalling channel. The signal current has been distinct in frequency from the supply current and gets filtered out in a relay fitted instead of a manual switch or time switch at each lamp (or lamp-group) position. The signal current gets injected into the network at the central point. Control could be achieved either directly from this point, or from some other position connected to the control point by a twin pilot cable.

R.M.S. Value : Root-mean-square value.

Robot : A mechanical flexible arm which is powered electrically, hydraulically or pneumatically, having several degrees of freedom to carry out a wide range of human-like movements controlled by computer and which are programmable to perform different functions.

Rochelle Salt : A piezo-electric cystal widely which is used for microphones and record player pick-ups. It is having a very high efficiency of conversion of mechanical to electrical energy and can be grown on a commercial scale.

Rod Electrode : An eath electrode. Copper rods of 12-18 mm diameter, cast iron pipes 10 cm or more in diameter and not less than 12 mm thickness, or galvanised steel water pipe not less than 40 mm diameter, have been normal.

Rod Gap : Refers to a form of spark gap. A simple rod gap has commonly used for the protection of power systems against lightning and other surge voltages. It consists of a pair of square-section metal rods of 12 mm side.

Roller-spot Welding : Refers to a variety of lap welding similar to seam welding but with the pulses of current so spaced that the spots produced have been each separate instead of overlapping.

Rolling-butt Contacts : Refers to the contacts in which the circuit gets interrupted and the arc formed at a part of the contact surface different from that used for conduction. Contact always gets established along a line. High-melting-point hard inserts could be provided at the tips of the contacts, and silver or silver alloy inserts can get fitted to the contact base where conduction occurs.

Rontgen : A unit of measurement (symbol : R) using the ionisation effect. It may be defined as the quantity of X-or gamma-radiation such that the associated corpuscular emission per 1.293 mg of air produces (in air) ions which are carrying 1 electrostatic unit of charge.

Rontgenogram : See radiograph.

Roof Conductor : Of a lightning protective system, a conductor connecting several air terminations, thereby extending the zone of protection.

Root-mean-square Value : This term refers to the square root of the average square of a variable, normally a time-dependent variable, over a specified interval. In particular, it has been the effective value of an alternating current or voltage; that is, the equivalent effect of the alternating current compared with a direct current in terms of a common property, such as heating.

Rosenberg Generator : Refers to a simple form of metadyne generator. It is having no interpoles, and the control windings surround poles which each carry the flux of the two adjacent polar projections.

Rosenberg Starting : A starting method for synchronous motors. A small auxiliary motor of the squirrel-cage, slip-ring or synchronous-induction type is having its

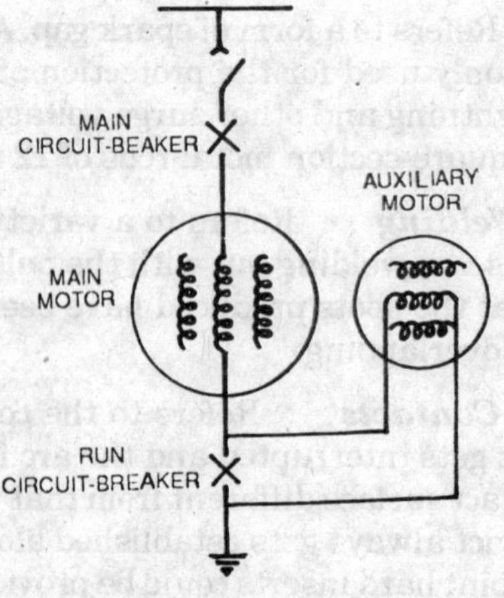

Fig. 4. Line diagram of Rosenberg starting scheme. The auxiliary motor is connected in series to reduce the current while starting.

stator windings connected in series with those of the main synchronous motor (see Fig. 4). On closing the main circuit-breaker with the run-circuit-breaker open, full voltage gets applied to the two motors in series, the resultant current getting reduced to a low value by the relatively high impedance of the auxiliary-motor windings. The auxiliary motor provides most of the torque required to reach slip speed, after which the main motor gets synchronised, the excitation adjusted and the auxiliary motor shorted out by the run circuit-breaker.

Ross Courtnery Eye : Refers to a cable eye or tag. The eye lies open for insertion of the wire and then gets bent over a washer to from a tag.

Rotary Converter : See converter, motor generator, motor converter, synchronous converter.

Rotary Substation : A substation, now obsolescent, having rotary equipment. Rotary substations were mostly those providing d.c. supplies, after transformation and conversion from high-voltage a.c. mains, to public supply districts or to traction networks through rotating machinery like synchronous converters or motor converters.

Rotary Transformer : Refers to a composite machine which is having a single magnet frams but two separate armature windings, one acting as a generator and the other as a motor, and independent commutators. It is also called a dynamotor.

Rotating Amplifier : Refers to a rotating d.c. genedrator whose function has been to convert mechanical to electric energy in such a way that the electric power output could

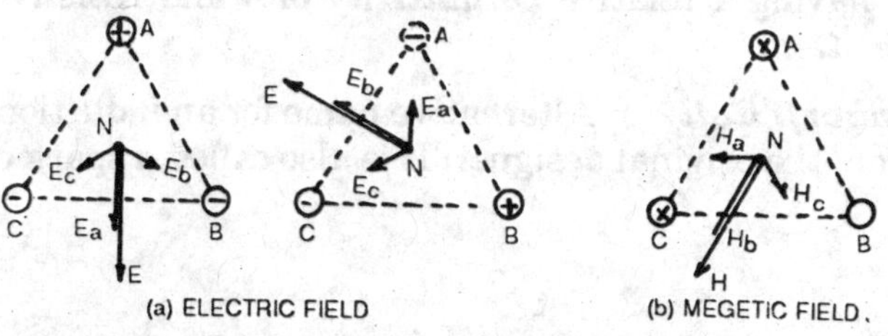

Fig. 5. Rotating fields of symmetrical three-phase system.

be accurately and rapidly controlled by a small electric signal applied to the control field of the machine. Rotating or rotary amplifiers find their main application in industrial closed-loop control systems.

Rotating Field : Refers to a vector field, the direction in space of which changes with time in a rotary manner. A system of M conductors having geometrical symmetry and to which symmetrical M-phase voltages have been applied to charge them or through which symmetrical M-phase currents flow, produces a rotating field, electric or magnetic respectively (see Fig. 5).

Rotational E.M.F. : Refers to an electromotive force which gets induced in a short-circuited coil of a rotating machine as a consequence of its cutting an air-gap flux in the commutating zone.

Rotor : Refers to the rotating component of a machine. The term is usually applied only to a.c. machines.

Rototrol : A type of single-stage rotating amplifer which is relying on the use of positive feedback.

Rousseau Diagram : Refers to means of determination of the luminous output of a light source from its polar curve. The average height of the Rousseau diagram provides the mean spherical candle power.

R.r.r.v. : Rate of rise of restriking voltage.

Rubber : Insulating material which is available in a natural or synthetic state. A cable insulation find use in various forms (tough-rubber sheath; vuicanised india rubber). As pure rubber it is having a relative permittivity of 2.6 and a volumer esistivity 10^{15} Ωcm; a vulcanised rubber is having a relative permittivity of 4 and resistivity of 10^{17} Ωcm.

Ruhmkorff Coil : Alternative name for an induction coil, from the original designer. It is also called a spark coil.

Sag : Of an overhead line, this term refers to the greatest vertical displacement of the line from the straight line between its points of suspension.

Sag-tension Relation : Of an overhead line, this term refers to the dependence of the sag on the tension of the line. A suspended conductor takes the form of a catenary. For all practical applications the form may be considered as a parabola. The relationship between sag, span and tension for a specified condition may be put as follows:

$$S = WL^2/8T$$

Where S=sag (metres), W=resultant load (newtons per metre), L=span(metres) and T-horizontal tension (newtons).

Salient-field Winding : Refers to a field coil around a salient pole a depicted in Fig. 1. This coil consists of a

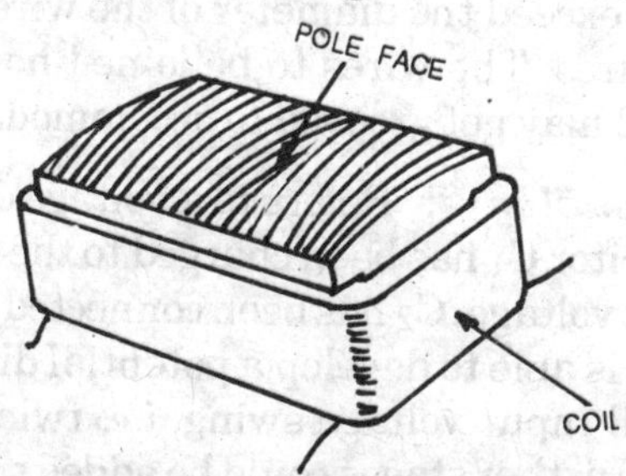

Fig. 1. Salient pole and salient-field winding.

number of turns (between one and several thousand) of wire or strip. Such windings find use for the stationary poles of all d.c. machines, for the rotating poles of low-and medium-speed synchronous machine, and for a few other specialised types.

Salient Pole : Refers to a pole piece which projects beyond the magnet yoke towards the armature.

Salient pole Alternator : Refers to an a.c. generator whose rotor field system gets incorporated salient poles projecting outwards from a hub.

Saturable-reactor : Deprecated term which was used in the past to signify a transductor.

Saturable-reactor Magnetometer : Refers to a magnato- meter in which a Permalloy or Mumetal wire gets subjected to high-frequency magnetisation (*e.g.* 5 kHz). Even-order harmonic fluxes get produced when a polarising field is present.

Sawtooth Waveform : Refers to the waveform in which the amplitude gets increased uniformly with time for a period then falls rapidly to zero in a comparatively short time.

S.C.A. : Steel-cored Aluminium.

Scanning : In television, this term is used to refer to the method which is used to resolve a picture into point-like elements for transmission as analogous electric signals, and reconstruction of the picture at the receiving end.

Scarf Joint : A joint used where the diameter of the joint must not exceed the diameter of the wire, as on overhead trolley wires. The wires to be joined have been cut obliquely and may get grooved to accommodate a locking pin.

Schenkel Doubler : Refers to a voltage doubler. In Fig. 2, the capacitor C_1 has been charged to the peak value of the a.c. input voltage. C_2 has been connected via two rectifiers so that it is able to develop a potential difference across it of the full input voltage swing, *i.e.* twice the peak input voltage. Further stages could be added so that the voltage may get trebled, quadrupled, etc.

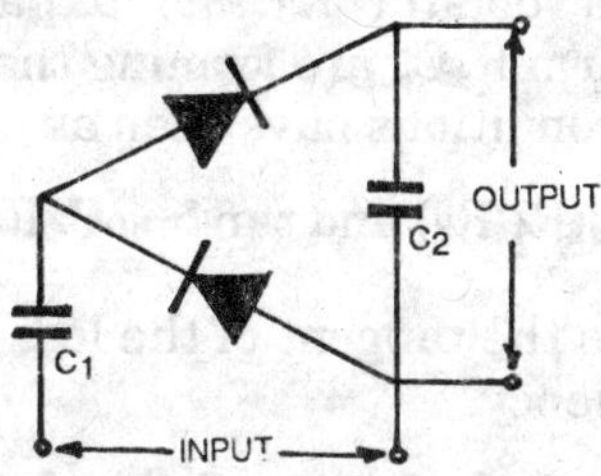

Fig. 2. Schenkel voltage doubler.

Scherbius Advancer : Refers to a phase advancer which resembles the Leblanc advancer with the addition of a stator winding in parallel with the armature winding (cf. Wlker advancer). The voltage across the stator winding has been proportional to the slip of the main induction motor, and its reactance has been also proportional to slip. It therefore takes a nearly constant current, thereby producing a constant flux, and generating a constant injection e.m.f.

Scherbius System : Refers to a system of motor speed control in which an induction motor is having its rotor connected to an a.c. commutator machine operating at slip frequency (Scherbius machine) and returning power to the line through an induction generator.

Schering Bridge : A standard equipment which is used for the measurement of capacitance and power factor of insulating materials. It has been of two different forms : (a) a high-voltage power-frequency bridge, and (b) a balanced audifrequency bridge. The h.v. bridge is shown in Fig. 3, in which an unknown cpacitor C_x to be measured and the standard capacitor C_s are forming the high-volt-

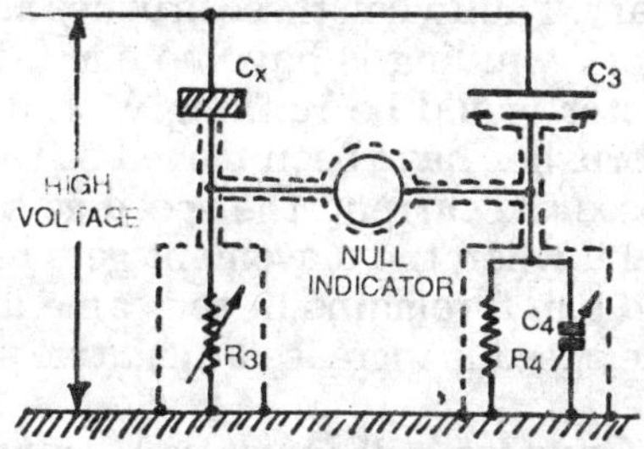

Fig. 3. Power-frequency high-voltage Schering bridge.

age arms: a resistor R_3 and a parallel resistance-capacitance arm R_4C_4 are forming the low-voltage arms. The balance conditions have been as follows:

$$C_x = C_s(R_4/R_3) \text{ and } \tan\delta = \omega C_4 R_4$$

where tan δ is the tangent of the loss angle and ω is the angular frequency.

Schrage Motor : Refers to a three-phase a.c. commutator machine which is having a shunt characteristic and used for variable-speed industrial derives. The primary winding (which is connected to the supply system) has been on the rotor and the secondary winding on the stator. The rotor is also carrying a low-voltage commutator winding, the conductors of which are located in the same slots as, and above, those of the primary windings. The brush gear consists of two movable rockers, each of which gets fitted with three brush spindles per pair of poles. The brushes attached to each rocker move over separate portions of the commutator surface to enable these brushes to be placed 'in line' or to be moved in either direction, so that more or less commutator segments have been included between a brush on one rocker and the corresponding brush on the other rocker.

As the primary winding has been located on the rotor, electromotive forces or slip frequency get induced in the secondary (stator) winding. The e.m.f.s. at the brushes have been also of this frequency. The e.m.f. induced in each coil of the commutator winding has been constant at all speeds, and therefore the e.m.f.s. injected, via the brushes, into the secondary winding have been proportional to the number of commutator segments included between corresponding brushes on the two rockers, *i.e.* between the brushes connected to a particular phase of the secondary. Thus when these brushes have been in line the secondary winding is have no e.m.f. injected into it and the motor would be running with its natural slip. When the brushes have been moved so that the injected e.m.f. opposes the current, the speed gets reduced (positive slip), and when the movement gets reversed so that the injected e.m.f. remains in the same direction as the current, the speed is increased (negative slip).

Scintillation Counter : Ionisation counter which consists of a phosphor which receives incident radiation and gives rise to light of a very low order, the light being detected and amplified by a photomultiplier followed by a main amplifier and a scaler or ratemeter.

S.C.M. : Symmetrically Cyclic Magnetic State.

Scott-bentley Discriminator : Refers to a single-step crane load discriminator.

Scott Connection : Refers to a three-phase/two-phase connection of transformers. Two single-phase units of different ratings have been needed but for interchangeability and the provision of spares it has been more usual to employ two identical units, each of which can be used in either position. The arrangement is depicted in Fig. 4.

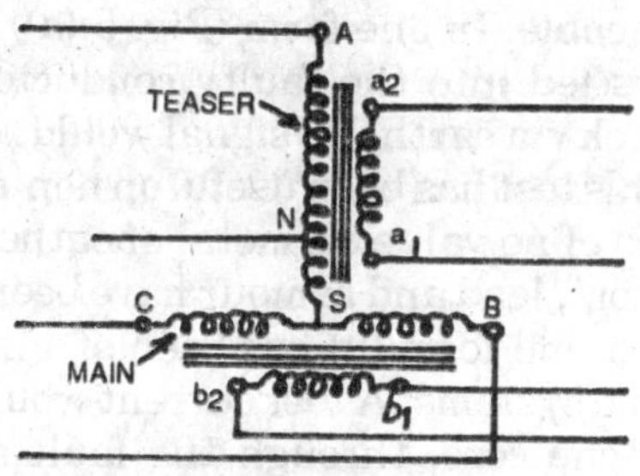

Fig. 4. Scott transformer connection.

S.C.R. : Short-circuit ratio.

S.C.R. : Silicon-controlled rectifier.

Screen-protected Machine : Refers to a machine in which all openings in the casing get covered with a wire mesh so that protection against accidental contact has been obtained without serious impairment of ventilation.

Screened Cable : A cable having the whole of the dielectric applied to each core before the cores get laid up. A metal screen has been applied over each core insulation, the screens being in contact with each other and the lead sheath.

Screening Reactor : Alternative name for line choke.

See Electrode : An electrode which has been immersed in the sea near the shore and connected to a land-line in an electric circuit where the sea has been used as a conductive path.

Sealing End : Refers to a closed box which is attached to the end of a cable where it connects to an external conductor, to protect the cable insulation from air or moisture. It is also known as a sealing box or sealing chamber.

S.C.S. : Silicon Controlled Switch.

Search Coil : Refers to a small inductor which is used for measuring magnetic field flux. It is also known as an exploring coil.

Search-coil Test : Refers to the direct method of locating a fault in a cable. In one form (Fig. 5 (a)) an interrupted current gets fed into the faulty conductor, through the fault and back via earth. No signal would be heard beyond the fault. This test has been useful on non-metal-sheathed cables but is of no value on metal-sheathed cables unless the conductors, lead and armour have been severed at the fault. The second form (b) has been of a greater value on modern cable systems. A test current would be allowed to pass along one core, through the fault and back via a second core. The resultant signal, although small, could be picked up between generator and fault with a coil connected to telephones and placed on the cable.

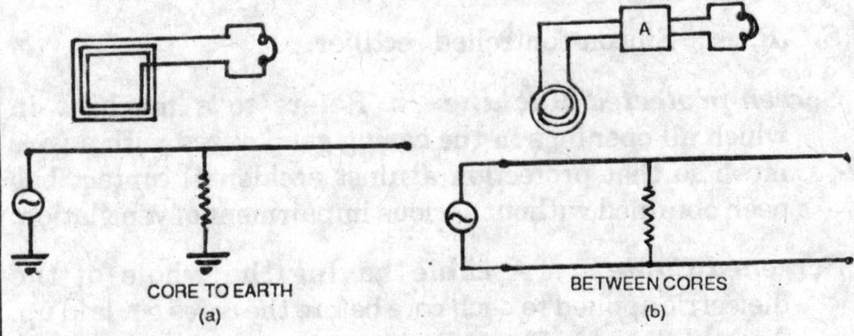

Fig. 5. Search-coil tests.

Secondary Cell : An alternative term for an accumulator.

Secondary Distribution : The term used for the low-voltage distribution from secondary substations to consumers' premises (cf. primary distribution).

Secondary Electrode : Alternative name for a bipolar electrode.

Secondary Emission : Refers to the emission of (secondary electrons from the surface of a solid when it has been bombarded by (primary) electrons or ions of sufficient kinetic energy.

Secondary Winding : Refers to the winding of a transformer from which the output is taken.

Section Gap : In an overhead conductor for electric traction, this term refers to an electrical and mechanical gap which is arranged to provide a continuous path for the current collector. This is achieved by overlapping the adjacent ends in the horizontal plane. It is also termed as an air gap or overlap span.

Section Insulator : In the overhead contact wire of a traction system this term refers to a device for dividing the wire elecrically into sections while maintaining mechanical continuity.

Sectionalised Busbar : Busbar which gets splitted into sections by using isolators or circuit-breaker and isolators. A single-busbar arrangement is depicted as a single-line diagram in Fig. 6. Incoming and outgoing circuits could then be shared between two sections so that

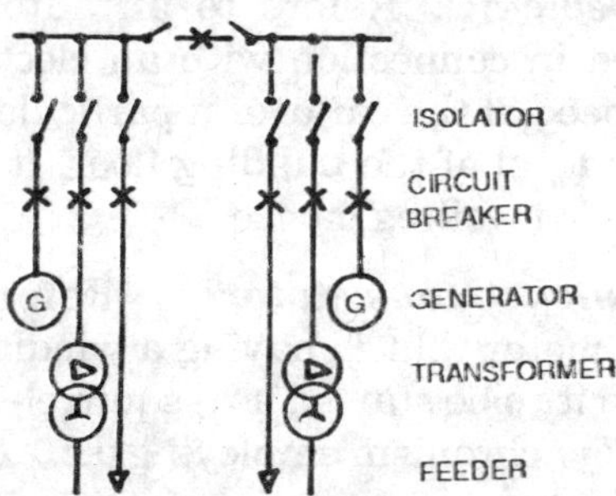

Fig. 6. Sectionalised single busbar, may provide better continuity of supplies and need simpler switchgear.

supplies may be maintained when a section of busbars have been shut down.

Seebeck Effect : Refers to a thermoelectric effect. If a closed electric circuit has been constructed of two (or more) dissimilar metallic conductors, and the junctions between different metals have been maintained at different temperatures, an electric current will flow in the circuit.

Selectivity : Refers to the ability of a circuit to respond more readily to signals of a particular frequency to which it has been tuned than to signals of other frequencies.

Selenium Rectifier : The most commonly used metal rectifier. The cells consist of a steel or aluminium plate, which gets coated with a thin layer of selenium to which suitable additions are made to improve the rectification. The coated plates undergo heat treatment and then get covered with a thin layer of an alloy which is having cadmimum, as this metal has been found to give the best rectification ratio. After the cells have been completed they have been electrically formed by passing a current through them in the reverse direction.

Self-excitation : Refers to the development by an electromagnetic machine of a working flux, without any external source of magnetising current.

Self Inductance : Refers to the ratio of the magnetic flux-linkage of an electric circuit and the current flowing in heat circuit.

Self-levelling Device : Refers to an automatic device which is used in connection with an electric lift car to reduce the speed of the car over a particular zone and to stop it at the level of the building floor. It is sometimes termed as a car-levelling device.

Self-starting Synchronous Motor : Refers to a synchronous motor that is having a winding in the pole faces to make it to be started as a squirrel-cage motor. It then runs in synchronism employing d.c. excitation.

Self Surge Impedance : More commonly termed simply as surge impedance.

Selsyn : Refers to a type of electric machine which is similar in many respects to a slip-ring induction motor but of special characteristics to make two or more units to rotate in mutual synchronism by using an electrical tie, analogous to an infinitely flexible shaft transmission of unlimited length.

Semiconductor : Refers to a crystalline material of high purity having a low conductivity, between that of metals and insulants. The principal charge carriers for current conduction have been free electrons in an n-type semiconductor and holes in a p-type semiconductor. Germanium and particularly silicon are semiconductors of major practical importance.

Semiconductor Rectifier : A device which uses the rectifying properties of a semiconductor junction.

Separator : Refers to the structure of insulating material used in an accumulator to separate plates of opposite polarity. It is normally formed of vertical rods or a diaphragm.

Sequence Switch : Another term for notching controller.

Series Capacitor : A power-system capacitor which gets installed at the isolating transformer for improving voltage regulation and reduce the kV A demand at source.

Series-characteristic Motor : Refers to a motor, the speed of which decreases with increasing load, *e.g.* a series -wound or heavily compounded-wound motor. It is also called an inverse-speed motor.

Series Motor : The term used for an a.c. commutator motor which are operating with series characteristic on single-phase supply.

Series/parallel Connection : The term used for the method of connection of electric apparatus in which (a) the items get connected alternatively in series or in parallel, or (b) some get connected in series and some in parallel.

Series/parallel Control : Of electric vehicles, this term refers to a means of control which involves the connection of d.c. motors, first in series and then in parallel.

Series/parallel Starter : Refers to a switching starter for two-phase induction motors. It has been connected so that the windings of each phase have been in series when in the starting position, and in two parallel circuits in the running position.

Series resonant Circuit : A resonant circuit having a single path with capacitance and inductance in series connection.

Series Transformer : Alternative term for current transformer.

Series Trip : A device which is used for releasing the restraining mechanism of a circuit-breaker, incorporating a tripping coil which gets energised by the main current passing through the circuit.

Series wound Motor : Refers to a d.c. motor in which the field and armature windings have been connected in series with respect to the supply (Fig. 7).

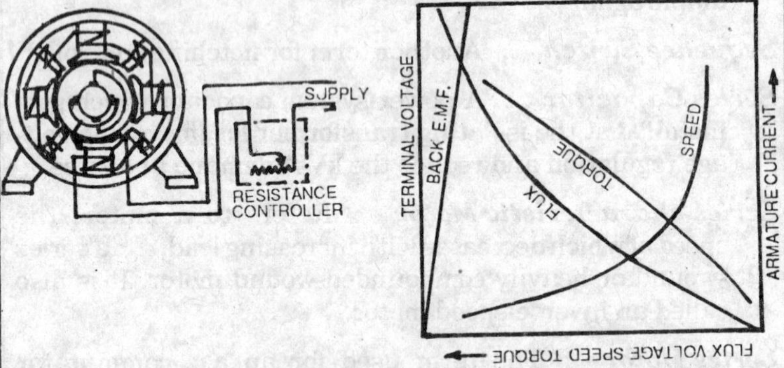

Fig. 7. Series wound Motor. Typical connections for winding and control gear are shown.

Serving : Of a cable, a layer of fibrous material, like jute, tape or yarn, which gets impregnated with a waterproof compound (*e.g.*.bitumen), and applied to protect the metal sheath or armouring.

Servomechanism : Refers to a closed-loop amplifying system having negative feedback, wherein the amplifier supplying the output has been activated by an error signal which has been derived from the difference between input and output.

Servomotor : A small motor which is used in a servomechanism. Most have been two-phase induction motors which are having a power output of $\frac{1}{2}$-100 W.

Shackle Insulator : Refers to an insulator, in an overhead- line system, supported by a pin which passed through it and gets attached at each end. It has been mainly used for low-voltage application, but has been manufactured for voltages up to 3.3 kV. It is also called a bobbin insulator.

Shaded Pole : The term used for an arrangement of a single-phase magnet to produce a 'shifting field' across its pole faces.

Shaded-pole Motor : Refers to a single-phase induction motor in which the phase-splitting of the stator flux has been provided by a permanently short-circuited auxiliary winding displaced in position from the main stator winding. The motor may be having salient poles as in Fig.8, in which case the auxiliary shading winding has been a bare copper band that encircles part of each pole face.

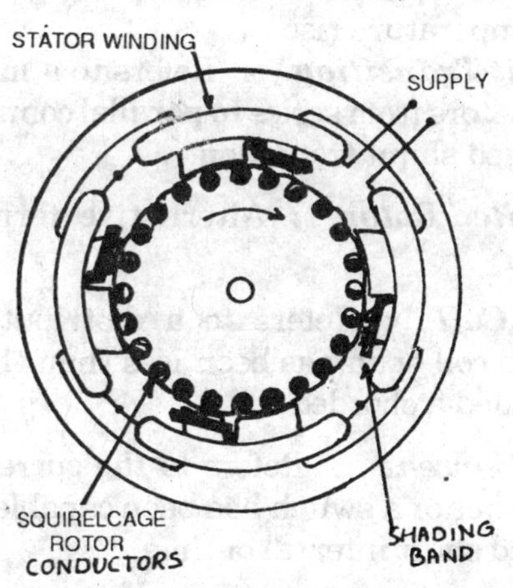

Fig. 8. Shaded-pole motor.

Shaft Cable : Cable for verticle installation, as a mine shaft. It has been a non-bleeding cable to disallow excess pressure of impregnating medium at the lower end.

Shell-type Transformer : Refers to a transformer in which the core laminations surround and generally enclose the windings.

Shock Discharge Test : The term used for a method of localising a fault in a cable by using the noise produced by a capacitor discharge through the fault.

Short-circuit, Short : Refers to a connection between two points of a circuit, especially across a source of electric energy, by a conducting path of low resistance.

Short-circuit Ratio : For an alternator running at rated frequency, this term refers to the ratio of the field excitation for rated voltage on open-circuit to the field excitation for rated armature current on short-circuit.

Short-circuit Testing :

(1) Refers to the testing of a circuit-breaker to prove its short-circuit rating. It has been an integral part of circuit-breaker manufacture.

(2) Refers to the testing of a generator or transformer having appropriate terminals short-circuited and full-load current passing to prove the design figures of temperature rise.

Short-circuit Transition : Refers to a method of charging d.c. motors from series to parallel connection, alternatively called shunt transition.

Short-circuited Rotor : Alternative term for squirrel-cage rotor.

Short-pitch Coil : Refers to a distributed winding in which the coil span has been less than the pole-pitch. It is also called a chorded coil.

Short-time Current : Refers to the current which a circuit- breaker or a switch has been capable of carrying for a specified short interval of time.

Shot Noise : Noise which occures in the anode circuit of a thermionic value as the current is not continuous but consists of a large number of random pulses taking place as each electron reaches the anode.

Shunt : Refers to the branch which gets connected in parallel with another branch of a network in particular, one so connected across an indicating instrument for the purpose of altering its sensitivity.

Shunt Capacitor : Refers to a power-system capacitor which gets installed on the l.v. side of distribution transformers or directly connected across single-phase motors (with the usual switching and other precautionary measures). Shunt capacitors give rise to an increase in the active power capacity of the line, a decrease in I^2R losses, and an improvement of voltage regulation.

Shunt-characteristic Motor : Refers to a motor, the speed of which remains more less constant as the load increases.

Shunt Limit Switch : See limit switch.

Shunt Transition : Refers to a method of changing d.c. motors from series to parallel connection, also called short-curcuit transition. One motor or set of motors gets short-circuited, then open-circuited and connected in parallel with the other motors.

Shunt Trip : A device which incorporates a tripping coil which is energised by a low-voltage supply. It is used to release the restraining mechanism of a circuit-breaker.

Shunt-wound Motor : Refers to a d.c. motor in which the field and armature windings connected in parallel with respect to the supply (Fig. 9).

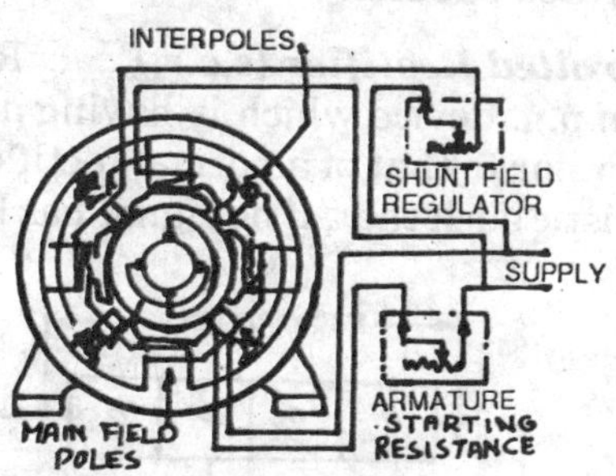

Fig. 9. Shunt-wound Motor. Typical connection for winding and control gear are shown.

Sideband : Means a band of frequencies which are extending above and below a carrier frequency, of total width equal to twice the highest modulating frequency.

Sidetone : Term used for the reproduction in the receiver of a telephone set of sounds, like the voice of the speaker, picked up by the microphone of the same set. It gets reduced by the use of hybrid coils.

Siemens : The SI unit of conductance and admittance (symbol: S).

Siemens Dynamometer : Refers to a dynamometer which has been designed for the measurement of current or power. The electromagnetic forces are balanced against the torison of a spiral spring.

Signal Control : Of a lift car, this refers two control method in which the car gets started from inside the car, but stopped by signals from any button, either on a landing or in the car.

Signal/Noise Ratio : Refers to ratio of the strength of a wanted signal to that of the noise interference present; it is generally expressed in decibels.

Silent Discharge : Refers to a noiseless high-voltage gas conduction which involves considerable energy.

Silica Gel : Desiccating agent which is commonly used in a transformer breather.

Silicon : Refers to a tetravalent semiconductor element which is widely used in the manufacture of semiconductor devices, especially diodes, transistors, f.e.t.s, thyristors and integrated circuits.

Silicon-controlled Rectifier (s.c.r.) : Refers to a four lay- er p.n.p.n. device which is having a reverse characteristic similar to that of a normal rectifier and a forward characteristic such that, if no signal has been applied to a

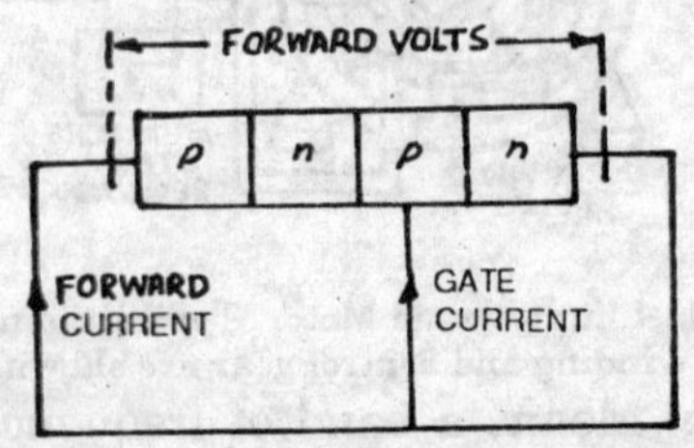

Fig. 10. Principle of a silicon-controlled rectifier.

trigger terminal, it will block positive 'anode' to 'cathode' voltages. If a signal has been applied to the trigger terminal, the device would be made to conduct and will go on doing so until the anode current gets reduced below a critical level,known as the holding current. (see Fig. 10). It is now called a thyristor.

Silicon controlled Swtich(s.c.s) : Refers to a four-layer device which is similar to a thyristor but with both gate leads brought out. It can therefore get switched on by a positive pulse on the cathode gate (as in a normal thyristor), or by a negative pulse on the anode gate.

Silicon Rectifier : Refers to a junction diode which is similar in construction and operation to a germanium rectifier, but with silicon instead of germanium.

Silicones : Refers to a semi-inorganic thermosetting plastics, which are being manufactured as fluids, greases, resins and rubbers. All are having extremely low-loss electrical insulating properties and in combination with mica, glass fibre or asbestos, will be forming a full range of insulating component that can be regarded for continuous application within the temperature range of -50° to +200°C and for intermittent operation upto 300°C.

Silmanal : Refers to a permanent-magnet material. It is an alloy of non ferromagnetic substances but have so large a coercive force that the magnet-strength gets unaffected by fields up to about 3000 A t/m.

Silver : A white metallic element having a specific resistivity of 1.62 μΩ per centimetre cube.

Simple Catenary Suspension : In electric traction, the turn refers to a form of constuction for overhead conduc-

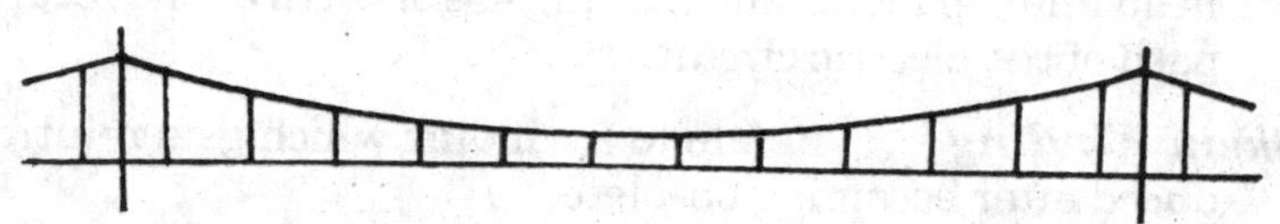

Fig. 11. Simple Catenary Suspension

tors in which the contact wire gets suspended by using droppers at intervals of $4\frac{1}{2} - 6$ m from a support wire which itself gets suspended as a catenary over the track (Fig. 11). The contact wire has been kept at a uniform height above the track by varying the length of the droppers.

Simplex Winding : Refers to an armature winding in which there has been one electrical path only for each pole.

Sine-wave Impedance : See impedance.

Single-break Circuit-breaker, Switch : Refers to a circ- uit-breaker of switch in which the circuit has been broken at one point only in each pole or phase.

Single-electrode System : Of an electrolytic cell, this term refers to one electrode and the portion of the electrolyte in contact with it. It is also called a half element or half cell.

Single-layer Winding : Refers to a winding having a single coil side in each slot.

Single-phasing : Refers to the effect in a three-phase system when two of the phases have been rendered effective; in particular the disconnection of one line of a polyphase induction motor when the machine has been running.

Single Potential : Deprecated alternative term form elect- rode voltage.

Single-way Connection : In a static convertor, this term refers to a connection of the converter devices such that the current in the transformer secondary winding gets passed only in one direction.

Single-wire Overhead Distribution : Refers to single-phase distribution system which is using only one overhead line conductor and making use of earth as the return path of the electric circuit.

Skein Winding : Refers to a winding which gets reintroduced after becoming obsolete.

Skin Effect : Refers to the phenomenon whereby, with alternating currents, particularly if the frequency has been high, the current carried by a conductor has been not uniformaly distributed over the available cross-section, but tend to get concentrated at the conductor surface. It has been due to magnetic flux that links part, not all, of the conductor.

S.L. Cable : A three-core cable which is hown in Fig.12, having each core sheathed with a single-lead cover.

Fig. 12. S.L. Cable.

Sleeve :

1. The term used for a short length of unthreaded tubing which is used to connect together the ends of lengths of plain conduit.
2. The term used for a casing of lead or copper which is kept round joints in cables and filled with compound.

Slip : Refers to the fractional speed-difference between a driving and a driven member, like a pair of pulleys connected by a belt, the two parts of a clutch, or the stator and rotor of an induction motor.

Slip Regulator : Refers to a device which is used for varying the slip of an induction motor rotor. It is generally used to decrease the motor speed when heavy loads occur.

Slip Resistance : Refers to a secondary resistance step which is incorporated in a motor controller to limit the current taken from the supply at the instant when the peak load gets applied to the motor in applications like press drives, guillotines, etc.

Slip Ring : Refers to a conducting ring which is connected with a winding and rotating with it, making an electrical connection to be maintained with an external circuit by means of a stationary brush resting on the ring.

Slip-ring Motor : An induction motor which has a wound rotor with the connections from this made to ship-rings.

Slow-break Switch : Refers to a switch in which the speed of operation of the contacts in breaking a circuit has been dependent entirely on the speed of action of the operator, *i.e.* no mechanical aid such as a spring has been included.

Slow Butt Welding : Alternative term for upset butt weld- ing.

Small Bayonet Cap. : Refers to a lamp cap which con- sists of a bayonet cap having a cap diameter of about 15 mm. It is frequently used for automobile lamps.

Small Centre Contact : Refers to a centre-contact cap which finds used for lamps and having a diameter of about 15 mm.

Small-oil-volume Circuit-breaker : Refers to a circuit-breaker in which are extinction occures in oil while air is used for insulation between phases, and sometimes to earth.

Smooth-conductor Cable : Refers to a cable in which a smooth surface has been provided for the conductor (as opposed to the ridged surface of a stranded or bunched conductor) by using a smooth metallic layer laid closely over the conductor below the dielectric.

Smoothing : The term used for the attentuation of ripple components, such as has been present in the output of low-power rectifiers, to have been a smooth direct-voltage or current output. It can be achieved by a capacitance-resistance filter circuit.

Snail Clamp : Refers to a variant of an anchor clamp which has been designed for use with heavy overhead-line cond- uctors. The conductor has been taken round a snail groove, the resultant static friction enabling a compara-tively small clamping plate to get used at the centre of the spiral.

Snap Switch : Another name for a quick make-and-break switch.

Snatch Off : The term used for the method of pulsing an automatic voltage regulator by a back-e.m.f. to reverse the force on the voltage sensing device.

Socket-outlet : A device which is used in domestic installations to facilitate the connection to the main supply of portable lighting fittings and other appliances. It is fixed and designed to receive a plug which is carrying protruding metal contacts corresponding to recessed contacts in the socket-outlet.

Sodium-vapour Lamp : A discharge lamp having sodium vapour, useful light being emitted from or excited by the discharge though this vapour.

Soft-iron Instrument : Indicating instrument which is better known as a moving-iron instrument.

Soil Warming : Heating of the soil for horticultural purposes by using transformer-fed low-voltage systems, with a grid of heating wires laid just below the surface of the ground.

Solar Cell : A device which is used for the direct coversion of solar energy to electric energy by a thermocouple or a photovoltaic cell. Powers between about 20 and 120 watts per square metre (*i.e.* efficiencies of 2 - 10%) could be obtained, and such devices are employed for supplying power to telephone repeaters located in isolated situations, and also for supplying space vehicles.

Solar Heater : A device which is used for the direct utilisation of heat radiation from the sun.

Solar Power Station : A power station which uses solar energy either directly by photovoltaic effect or indirectly by thermal transformation.

Solder-pot Relay : Refers to a form of thermal relay, which is frequently fitted in motor control gear.

Solenoid : Refers to a coil of insulated wire which is intended for connection in an electric circuit to produce a concentrated magnetic-field. Solenoids are designed for operating on either direct or alternating current. D.C. solenoids are having a solid iron construction with a solid iron plunger working in a close-fitting brass tube. A.C. solenoids are having laminated cores to reduce eddy-current loss, although some smaller a.c. solenoids have a solid iron construction. Single-phase solenoids invariably have a shading ring.

Solid Carbon : A homogeneous carbon rod which is used as an electrode in a carbon arc lamp.

Solid-core Insulator : See post insulator.

Solid End : Of a cable, this form refers to an end which is hermetically sealed, with all the conductors sweated to a lead cap at the end of the sheath.

Solidal Cable : Trade name for a cable which is having solid, shaped conductors of soft aluminium.

Solidly Earthed : Connected effectively to earth without intervention of any device like a fuse, swtich, circuit-breaker, resistor, reactor or solenoid.

Sound Level Meter : An elaborate electric instrument which is used to measure noise levels.

Source Current : Refers to the current I_o which is supplied by the active element in the presentation of a source of electric energy as an equivalent current generator.

Source Voltage : Refers to the voltage E_o across the terminals of the active element in the presentation of a source of electric energy as an equivalent voltage generator.

Space Factor : Of bunched cables, this term refers to the ratio of the sum of the effective overall cross-sectional areas of the cables to the internal cross-sectional area of the conduit, duct, trunking, etc. in which they have been installed. The effective overall cross-sectional area of a non-circular cable has been taken as that of a circle of diameter eqaual to the major axis of the cable.

Span-length : Of an overhead conductor, this term refers to the horizontal distance between two adjacent points of support.

Spark : Refers to a short-duration current between electrodes in a gas, which is accompanied by the production of light and heat in the spark path.

Spark Coil : Alternative name for induction coil and Ruhmkorff coil.

Spark Gap : A pair of electrodes have been so designed that a spark or an arc could safely pass between them when the voltage across exceeds the breakdown value. The chief purposes of such gaps have been used for the measur- ement of high voltages and for protection of apparatus against damage due to excessive voltages. Common types of spark gap have been the sphere gap, needle gap, rod gap, and horn gap.

Sparking Plug : A device which finds use in automobile electrical engineering. It is having a spark gap connected between earth or the chassis of a vehicle and an insulated central electrode supplied from the h.t. distributor. It has been screwed into the cylinder head of a petrol engine to provide ignition.

Specific Inductive Capacity : Obsolete term for relative permittivity.

Specific Loading : In electric machine design, this term refers to the mean flux desnity over the air-gap surface where the electromotive force is induced. The specific electric loading refers to the number of ampere-conductors (product of number of conductors and the r.m.s. current per conductor) for unit length of armature or stator periphery.

Specific Resistance : Obsolete term for resistivity.

Sphere Gap : A spark gap which is having spherical electrodes of brass, copper or aluminium. It has been widely used as a sub-standard voltage-measuring device for voltages between about 2 kV and 2.5 MV.

Split Fitting : Refers to an angled conduit connection fitting which gets splitted longitudinally and held together by screws. It can be kept in position after the wires have been drawn into the conduit.

Split-phase Motor : it is a single-phase induction motor kept in fitted which an auxiliary stator winding dispaced in magnetic position form, and connected in parallel with, the main stator winding. The circumferential displacement of the two stator fluxes could be obtained by fitting the coils of the auxiliary winding midway between the main stator coils.

Splitter : Refers to a consumer's distribution unit which is having distribution fuses and main switch. It is suitable where the total connected load is not in excess of about 60 A.

Spot Welding : Refers to a variety of lap welding which is used generally in place of rivets in joining sheet metal or pressings, the current being concentrated at the weld by using electrode tips that have limited area.

Spring Drive : Alternative term for quill drive.

Spur : Refers to an extension from a ring, circuit. When it has been not convenient to include all the sockets on the ring, a limited number could be connected as spurs. Such spurs, having not more than two sockets, may be looped either from a socket on the main ring or from a junction box.

Square-loop Characteristic : Refers to a hysteresis loop for a mterial such that an abrupt change in response takes place at a particular flux density. Such a characteristic, for moderate applied field intensities, gets saturated in one direction or the other, giving a two-state property useful in magnetic amplifiers and computers.

Squirrel-cage Motor : Refers to an induction motor which has a rotor with a squirrel-cage winding. It has a poorer starting performance in terms of the ratio of starting torque to starting current than a slip-ring motor.

Squirrel-cage Rotor : Rotor of an induction motor having on it a squirrel-cage winding. It is also called a cage rotor or short-circuited rotor.

Squirrel-cage Winding : Refers to a type of rotor winding which is commonly used in an induction motor. A series of bars gets accommodated in the rotor slots, all bars being connected at each end to a common conducting ring. The bars and end-rings then are forming a construction resembling a cage. This cage may be considered as a superimposed series of full-pitch turns formed by pairs of bars a pole-pitch apart, joined together into a closed loop by the end-rings. If there has been an adequate number of bars, the cage winding accommodates itself to any number of poles on the stator. It is often called a cage winding.

Stabiliser :

(1) A device which is included in a circuit to maintain a constant potential difference between two points.

(2) Equipment which is used to redue automatically the rolling motion of a ship. It consists of fins extending out below the water line and tilted by an electrohydraulic servo-mechanism.

Stabilising Winding : Refers to a delta-connected winding which is used on star/star-connected transformers, or auto-transformers, to facilitate the flow of zero-phase-sequence currents, reduce third-harmonic voltages, reduce interference due to third-harmonic currents in the lines and earth, and stabilise the neutral point of the fundamental frequency voltage.

Stability : The term used for the ability of a system to return to a normal condition after being subjected to a disturbance. On a power system under normal conditions the various machines would be running in synchronism with each other, the relative angular positions of their rotors being found out by the power transfers between them.

Stamping : Of a machine or transformer this term is used for each of the thin, metallic sheets which form part of the core. The sheets are also known as core plates, laminations, or punchings.

Standard Cell : It is a special form of primary cell that will produce a constant electromotive force under varying conditions of use for a long period, provided that only a very small current has been drawn.

Standard Ohm : Obsolete term for ohm.

Standard Solenoid : A solenoid which is used with a search coil to form a standard of magnetic flux linkage.

Star Connection : The term used for the arrangement of phase windings into a group by connecting to a star point, or common termincal, all the corresponding ends of the phase windings. Normally the phase voltages have been different only in time phase in a symmetrical manner. If the phase voltages have been

$V_1 = V\angle O$

$V_2 = V\angle -(2\pi/m)$

....

$V_m = V\angle -(m-1)(2\pi/m)$

then the voltage AB in Fig. 13 would be V_1-V_2, the voltage AC would be V_1-V_3, etc. The current leaving line A has been the same as the current in phase 1, etc.

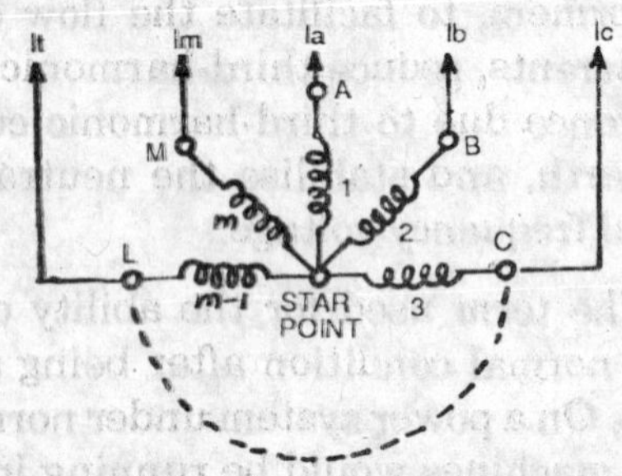

Fig. 13. Star connection.

Star/Delta Starting : Refers to a method of reducing starting current of three-phase motors. It finds use in conjunction with a motor designed to operate with the primary winding connected in delta, but with six terminals brought out for connection in star during starting. The starting current and torque have been one-third the value obtained with direct-on-line starting.

Star/mesh Conversion : Refers to a method for the simplification of networks which are operating at a given single frequency. In star/delta (three-branch) conversion, a star-connected network of impedances Z_a, Z_b, Z_c con-

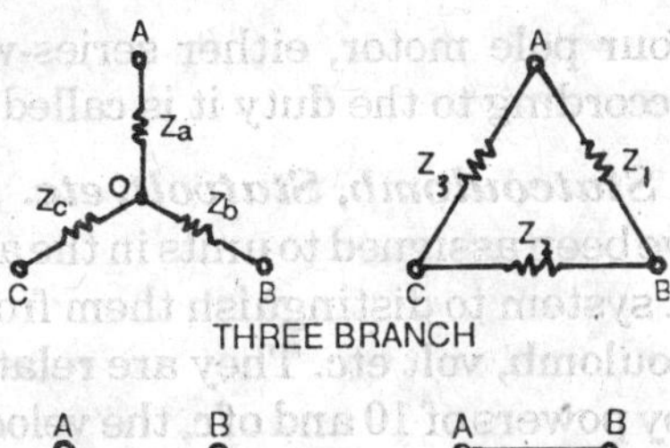

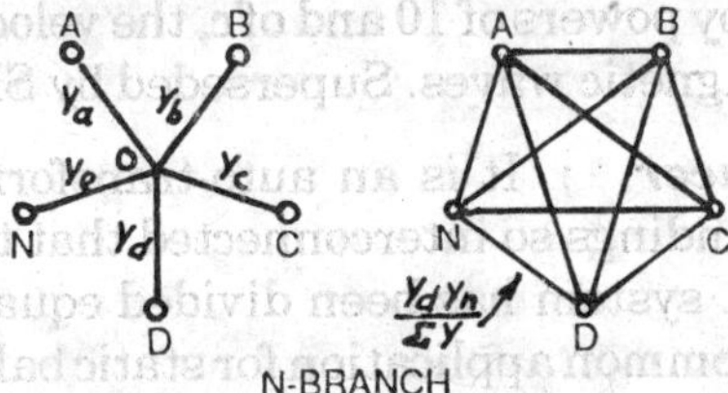

Fig. 14. Star/mesh conversion.

nected between terminals ABC in a network (Fig. 14) have been replaced by a delta connection of Z_1, Z_2, and Z_3 between terminals AB, BC and CA respectively, where

$$Z_1 = Z_a + Z_b + Z_a Z_b / Z_c$$

$$Z_2 = Z_b + Z_c + Z_b Z_c / Z_a$$

$$Z_3 = Z_c + Z_a + Z_c Z_a / Z_b$$

A delta connection could be converted to a star when $Z_a = Z_3 Z_1 \phi Z$; $Z_b = Z_1/Z_2$; and $Z_c = Z_2 Z_3 / Z$ (where $Z = Z_1 + Z_2 + Z_3$). The star and delta have been then equivalent in the sense that the impedance measured between any pair of terminals is the same for each.

Star Point : Refers to the point at which the branches of a winding in star connection have been joined together, and sometimes to earth.

Star Voltage : Refers to the voltage between any line of a three-phase or six-phase system and the neutral point of the system. An unsymmetrical system may be having more than one value of star voltage. It is also called Y-voltage or phase-voltage.

Starter Motor : In automobile electrical engineering, this term refers to a motor which is having a small opinion mounted on an extension of its armature shaft which engages with a gear ring on the periphery of the flywheel to rotate the latter. The motor has been usually a straight-

forward four pole motor, either series-wound or series-parallel according to the duty it is called on to perform.

Statampere, Statcoulomb, Statvolt, etc. : Names which have been assigned to units in the absolute electrostatic c.g.s. system to distinguish them from the 'practical' ampere, coulomb, volt etc. They are related to the practical units by powers of 10 and of c, the velocity of free-space electromagnetic waves. Superseded by SI units.

Static Balancer : It is an auto-transformer or inductor having windings so interconnected that the voltage of an a.c. or d.c. system has been divided equally between the wires. A common application for static balancers has been to convert a supply from two-to three wire, or from three-to four-wire.

Static Converter : Refers to a converter which is based on electronic devices of the semiconductor, mercury-arc or gaseous type, generally in combination with a transformer.

Static Electrification : Refers to a surface contact potential difference that exists across the boundary when two bodies of different materials are in intimate contact. The potential has been of the order of a fraction of a volt, but if one (or both) of the materials has been a non- conductor, the charges will get retained when the surfaces are again separated.

Static Machine : Alternative name for electrostatic generator.

Static Regulator : An automatic voltage regulator that is not having moving parts. It usually employs a magnetic amplifier.

This arrangement ensures suitable operation, even under condition of system short-circuit, or the equivalent.

Static Relay : Refers to a rectifier device which can be used as a relay without moving parts. Its action is dependent on the fact that the impedance presented to a small alternating voltage may be controlled by the value and direction of a steady direct voltage on which the alternating voltage is superimposed. Examples are the thermionic valve which is operated at grid voltages below and above cut-off, and its counterpart in transister circu- itry.

Static Susbstation : A term formerly applied to substations on d.c. networks equipped with static transformers, to distinguish them from rotary substations.

Stationary Battery : Refers to an assembly of accumulators which are erected on a fixed site and not intended to be moved.

Statistical Lag : In a spark gap, this term refers to the time between the application of a breakdown voltage and the chance occurrance of an electron in a position suitable to initiate breakdown.

Stator : The term used for the fixed or stationary component of a machine. The term has been generally applicable only to the stationary magnetic parts of a a.c. machines and the associated windings.

Stator-fed Shunt Motor : Refers to an a.c. commutator motor which is having characteristics similar to a schrage motor. The primary winding has been on the stator as in a normal induction motor, there have been no slip rings, and speed variation has been adjusted by means of an induction regulator connected between the brush gear and the line.

Steady-arm, Steady-brace : A fitting which is used with catenary suspension to keep the contact wire in its correct lateral position.

Steady-state Characteristic : Of an arc, this term refers to the relation between the voltage gradient and the current flowing through the arc.

Steady-state Stability : Stability which follows small or gradual disturbances, such as a slow increase in load on a power system.

Steam Turbine : A device which is for converting the potential energy stored in steam under pressure into mechanical work in a form suitable for driving electric generators, compressors or other machinery. The conversion is accomplished in two stages : first, kinetic energy is obtained by allowing the steam to expand, and secondly the resultant jets of high-velocity steam impinge on blades which have been forced to rotate in the same direction as the steam jets, so that power gets transferred to a shaft.

Steel : It is an alloy of iron having less than 2% of carbon and smaller amounts of manganese, silicon, phosphorus, sulphur and oxygen. It finds use as a conductor when mechanical strength is important. It is also used, as a permanent-magnet material, and for cases for transformers, electromagnets; etc.

Steel alkaline Cell : Refers to a storage cell or accumulator which uses an alkaline electrolyte. Its two distinct types, the nickel-iron cell and the nickel-cadmium cell are known; both of which use an electrolyte of dilute potassium hydroxide (casustic potash) and have a positive plate of nickel hydrate. In both types the plates are assemblies of steel tubes or pockets perforated minutely all over, enclosing the active materials. The average voltage per cell on discharge at 5-10 hr rate has been approximately 1.20 V.

Steel-cored Aluminium : Conductive material which has layers of aluminium wire encircling a core of galvanised steel strands. It is commonly abbreviated s.c.a.

Steel-tank Rectifier : Refers to the type of mercury-arc rectifier in which the arc occures in a steel tank.

Stitched Catenary Suspension : In electric traction this turn refers to a form of construction for overhead cond- uctors which is being used as an alternative to compound catenary suspension. It is similar to simple catenary suspension but the contact wire near a support structure gets held by an additional wire attached to the catenary as shown in Fig.15.

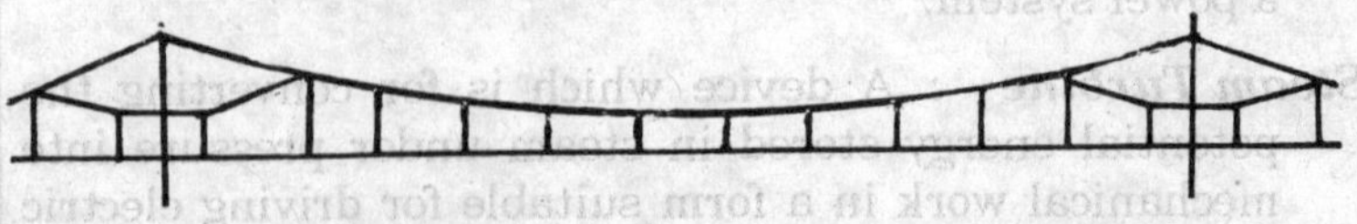

Fig. 15. Stitched Catenary Suspension

Stockbridge Damper : Refers to the most common vibration damper which is used to prevent the vibration of an overhead line. It is having two weights at the end of a short length of stranded steel cables suspended from the conductor at a point midway between two nodes of a vibration, where the amplitude would be a maximum.

Storage Cell : An alternative term for an accumulator.

Straight-through Joint : It is a form of cable joint in which the ends of two abutting cables have been joined to form one continuous length.

Strain Gauge : A device which is used primarily for detec- ting and measuring small variation in the dimensions of the surface to which it gets attached. The pick-up is able to convert mechanical movement into change in an electric quantity, the variations of which get subsequently amplified an impressed on a recorder.

Strain Insulator : Refers to an insulator, in a overhead-line system, which has been capable of transmitting the tension of the conductor to the supporting tower. It is also called a tension insulator.

Stranded Conductor : A conductor which is composed of several wires twisted together. The direction of twist of adjacent layers gets reversed.

Stray Load Loss : Refers to the additional loss in an electric machine or a transformer. It is caused by the load current and due to changes in flux distribution and to eddy currents.

Striking Voltage : Deprecated term for ignition voltage.

String Electrometer : It is a form of electrometer in which a metallised quartz fibre gets hung between fixed plate conductors, held at a high potential differences (Fig.16). A small test voltage applied to the fibre makes it to move towards one or other fixed plates. The deflection is observed by means of a microscope.

Stringing :

1. Running and tensioning the conductors on the supports of an overhead line.
2. Assembling suspension insulators into units or

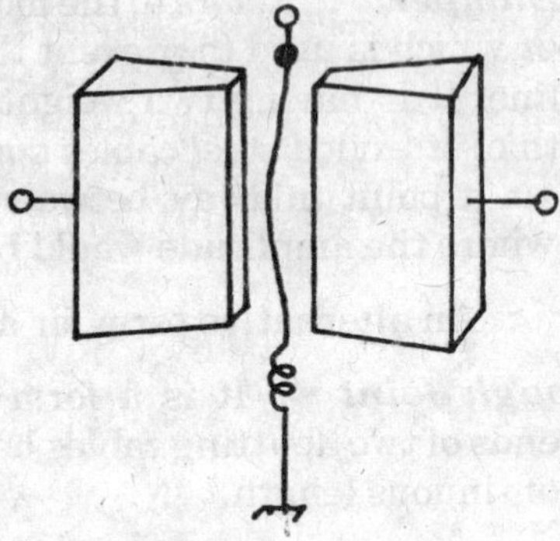

Fig. 16. String electrometer.

strings.

Strip Electrode : Refers to a form of earth electrode. Untinned copper strip about $25\times1\frac{1}{2}$mm has been normal, but scrap lengths of bare stranded copper have been also suitable.

Stubbs-Perry System : Refers to a method of load equalisation which is similar to Ilgner system. It is used when the generators are driven by steam turbines.

Submerged Arc Welding : The term used for a form of consumable-electrode arc welding that uses a bare wire in conjunction with an over-burden of granulated flux. The arc has been submerged beneath the flux, part of which fuses to form a slag, the unfused portion being recovered and re-used.

Submersible Motor : A machine which is used for driving underwater and borehole pumps. The dry type need a chamber intermediate between the pump and the motor. The wet type is submersible in the litral sense.

Substation : Refers to an assemblage of equipment which includes any necessary housing, for the conversion, transformation or control of electric power. It is customary now to classify substation according to the type of system within which a particular substation is included, and to use a qualifying prefix,for example transmission substation, 400 kV substation, distribution substation, 33 kV substation, converter substation, traction substation.

Sub-transient Reactance : Refers to reactance associated with the analysis of a synchronous machine that deter- mine the maximum fault current under symmetrical-fault conditions. It is corresponding to the leakage flux that occurs in the initial stage of a short-circuit.

Sulphur Hexafluoride : A heavy, very stable, colourless, odourless, non-toxic, non-flammable electronegative gas which is used in high-voltage circuit-breakers.

Superconductive Device : Refers to a device such as the cryotron, which is dependent for its action on the effect of a magnetic field on the critical (or transition) temperature below which certain materials show superconductivity.

Superconductivity : Refers to a phenomenon which is shown by materials at very low temperatures. At absolute zero (OK=-273.2°C) the atoms cease to vibrate and free electrons can pass through the lattice possessing little hindrance. At about 5K the resistance of certain metals becomes precisely zero, so that a current, once started, may persist for a long time due to its associated magnetic energy storage and complete absence of resistance dissipation without any applied electric field. This phenomenon could not be satisfactorily explained.

Superexcitation : Refers to the application of the principle of quick-response excitation to synchronous machines. It gives rise rates of voltage build-up of the order of 6 or 7kV/s. Exciters designed for superexcitation have been of high rated voltage (*e.g.* 600 V for a 250 V excitation) and have a correspondingly high ceiling voltage, approximately 1000 V.

Supegritirid : Transmission lines which are incorporated in the grid system and rated at 275 kV and above in many countries.

Superposion : A principle which is used in network analysis. The effect on, or the reponse of, any linear system due to the simultaneous action of a number of impressed causes or disturbances could be known by considering the effect due to each cause taken separately and summing them. When it is applied to an electric network, it implies that the current produced in various branches by a number of sources have been obtainable by superposing the currents resulting from the generators acting separately.

Supplementary Anode : In electroplating it is a small anode that is provided near deeply recessed parts of a cathode surface for facilitating a uniform deposition.

Suppressed-zero Instrument : Refers to a measuring instrument that does not get deflected until the deflecting force exceeds a preset minimum value. It is also called a set-up scale instrument.

Surface-barrier transistor : A transistor which is produced by an electrolytic process, having a very thin base layer and small junction area. It is operable up to about 70 MHz.

Surface Resistivity : Refers to the resistance between the opposite edges of a square of unit dimensions on the surface of insulating material. The value does not depend upon the size of the square, provided that the surface is having uniform properties: the resistivity is therefore commonly expressed in MΩ 'per square'.

Surge : A transient overvoltage on an electric network, usually in the form of a travelling wave on a transmission line. In such a case the surge may be initiated by initiated strokes or switching operations.

Surge Absorber : Refers to a protective device which is connected in series with an overhead line at substation terminals for absorbing some of the energy of a surge voltage.

Surge-current Indicator : Alternative name for a magnetic link.

Surge Diverter : Refers to a device which is connected between a transmission line and earth so as to divert to earth a momentary high-voltage surge. It normally comprises one or more gaps in series with non-linear resistors.

Thyratrons, ignitrons, and cold-cathode electronic devices may be used in some situations as surge diverters.

Surge Generator : A device which is used to produced high-voltage or heavy-current surges. It usually consists of a number of cpacitors that have been charged in parallel. For a high-voltage surge they have been discharged in parallel.

Surge-impedance : Refers to the ratio voltage/current in a single surge travelling in one direction/on a transmission line. For aline of inductance L and capacitance C per unit length, the surge impedance in ohms have been $Z_o=\sqrt{(L/C)}$. This is actually the self surge impedance. The mutual surge impedance refers to the voltage in a surge along one line to the current in the associated ratio of the surge along another line.

Surge-limiting Electrolytic Capacitor : An electrolytic capacitor which is designed to restrict the maximum voltage which could be maintained across its terminals.

Surge-proof Electrolytic Capacitor : Refers to an electr- olytic capacitor which has been designed to withstand momentary or intermittent surge of voltage in excess of the rated working voltage.

Surge Ratio : The term used for the ratio of the break-down voltage of some insulant material when a surge voltage has been applied, to that when a sustained 50 Hz voltage has been applied.

Surge Test : Refers to the application of a single voltage pulse obtained from a surge generator. Such a test has been specified by the peak value of the voltage wave, the time in microseconds taken to reach 90% of peak value from 10% of that value and the time in microseconds for the voltage to get decayed to 50% of peak voltage.

Susceptance : Refers to that part of admittance concerned with reactive elements. An admittance $Y = G + jB$ has been composed of two branches in parallel, one of conductance G, concerned with dissipation, and the other of susceptance B.

Susceptibility : In a magnetic circuit, this term refers to the ratio of the intensity of magnetisation to the magnetic field strength. It is also called volume susceptibility; symbol K.

Suspension Insulator : Refers to an insulator, in an overhead-line system that hangs freely from the cross-arm for supporting the line conductor. It consists of a number of separate units which are linked to form a flexible string.

Swan-neck Insulator : A pin insulator which is having the pin shaped to bring the insulator into the same horizontal plan as the support.

SWG. : See standard wire gauge, wire gauge.

Swtich : The term used for a mechanical or electrical device which is used for making or breaking, non-automatically the load current in a circuit.

Switch-fuse : Refers to a combination of a switch and one or more fuses. The fuses would not get carried on the moving part of the switch as in the fuse-switch.

Switch-type Voltage Regulator : Refers to an electromagnetic device which is having a primary winding in parallel and a secondary winding in series. The voltage impressed on a load could be adjusted by varying the number of turns in one or both of the windings.

Switchboard : Refers to an assembly of switchgear.

Switchgear : it is an apparatus which is used for controlling the distribution of electric energy or for controlling or protecting apparatus connected to a supply of electricity. The main components have been switches and circuit-breakers with ancillary apparatus like current and voltage transformers and cable-sealing ends.

Switching Station : A substation that is having associated switchgear and busbars.

Switching Transient : Refers to a transient which is impo- sed on a circuit by the operation of a switch. The closing of a switch has been equivalent to the injection across the switch terminals of a voltage equal and opposite to that existing across the switch when open.

Symmetric Circuit Element : Refers to a circuit element having two terminals whose voltage/current characteristic which is quantitatively independent of the polarity of voltage and of the direction of current.

Symmetrical Breaking Capacity : Refers to the r.m.s. value of the a.c. component of current which can be broken at a stated voltage simultaneously by all poles of a circuit-breaker.

Symmetrically Cycle Magnetic State : The term used for a condition of a magnetic material when it has been in a cyclically magnetised state, and the limits of the applied magnetising forces have been equal and of opposite sign, so that the limits of flux density have been also equal and of opposite sign. It is abbreviated s.c.m.

Synchro : It is a small instrument, up to about 100 mm diameter which is built like an induction motor and used for the transmission and reception of both electrical and position data. The basic synchro consists of a symmetrical three-phase stator and a single-phase rotor. A voltage is generated in each stator winding of value determined by its axis relative to that of the rotor winding.

Synchrocyclotron : The term used for an orbital accelerator in which the relativistic energy limitation of the synchrotron is removed, by modulating the frequency of the oscillator.

Synchroniser : It is a dial-and-pointer instrument which is used for synchronising an incoming machine. The pointer shaft is carrying a rotor acted upon the magnetic field of two coils, energised respectively by the machine and the busbar voltages. One form is shown in Fig. 17.

Synchronising : The term used for the process of connecting to a.c. supplies together in parallel; the selection of the appropriate instant for switching a synchronous a.c. generator on to energisted busbars or into parallel with another, normally running, synchronous machine. Before a generator could safely get connected to live busbars it becomes necessary for the voltage and frequency to be approximately the same as that of the busbars. In the case of a polyphase machine, the phase sequence also should

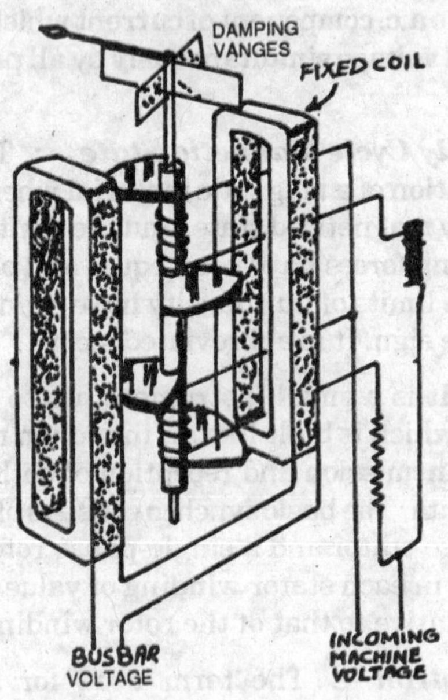

Fig. 17. Synchroniser

be indentical.

Synchronism : Refers to the condition that two separate periodic operations have been in step with one another.

Synchronoscope : An instrument which finds use in the process of synchronising two a.c. generators or systems, it shows the instantaneous phase difference between the two separate voltages.

Synchronous Clock System : Refers to a system in which clocks are operated from a.c. mains.

Synchronous Compensator : Refers to a machine which has been connected to a supply system to provide reactive MV A, normally leading but sometimes also lagging, without at the same time furnishing mechanical power to reactive MV A, normally leading but sometimes also lagging, without at the same time furnishing mechanical power to external plant. The output of the unit has been adjusted, both in sign and magnitude, by alteration of of the exciting current, invariably under the control of an automatic regulator. The primary use has been in the correction of power factor and in the improvement of transmission line stability.

Synchronous Generator : Refers to an a.c. generator or alternator that has been driven at a constant speed corresponding to the output frequency required.

Synchronous Impedance : Refers to the ratio of the short-circuit field current to the open-circuit current of a synchronous machine under specified conditions. It is generally expressed as a percentage.

Synchronous Convertor · Refers to a form of converter in which the armature is having connections to slip-rings for the a.c. input (two for single-phase and three or six for three-phase) and to a commutator for the d.c. output. As the circuits have been electrically connected, the voltage ratio has been fixed and a transformer has been usually necessary at the input end. Excitation can be provided from the d.c. output or from an exciter. Fig. 18 depicts the normal connections of a six-ring synchronous converter.

Synchronous-induction Motor : It has been a compromise between the slip ring motor and the synchronous motor. The good starting performance of the induction motor has been combined with the high running efficiency and variable power factor of the synchronous motor. It has been started as a slip-ring motor torque has been provided by the interaction of a rotating magnetic field produced by the primary windings and the poly-phase currents induced by this field in the rotor windings. When full speed as an induction motor gets reached d.c. excitation would be applied to the secondary and the rotor pulls into synchronous speed. The required power factor of the load would be then obtained by adjustment of the field-excitation current.

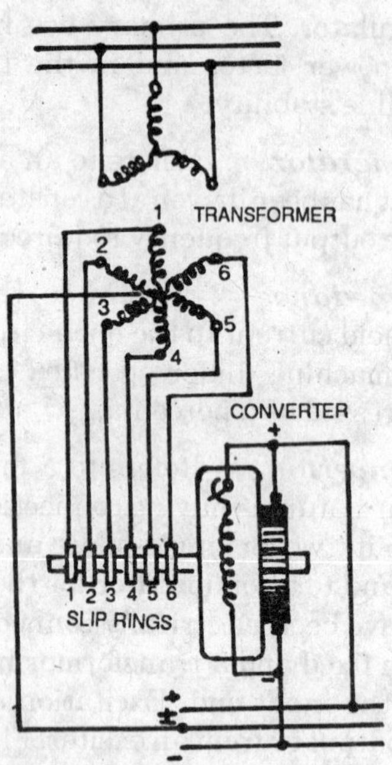

Fig. 18. Diamettrally-connected six-ring synchronous convertor.

Synchronous Motor : Refers to a constant-speed machine. Its speed gets dependent upon the frequency of the alternating-current supply to which is connected, and the number of poles for which it is designed (Fig. S 28).

Synchronous Reactance : The term used for the complexor difference between the synchronous impedance of a synchronous machine and its effective armatures resistance.

Synchrotron : An orbital accelerator that got developed from the betatron.

Synthesis : See network synthesis.

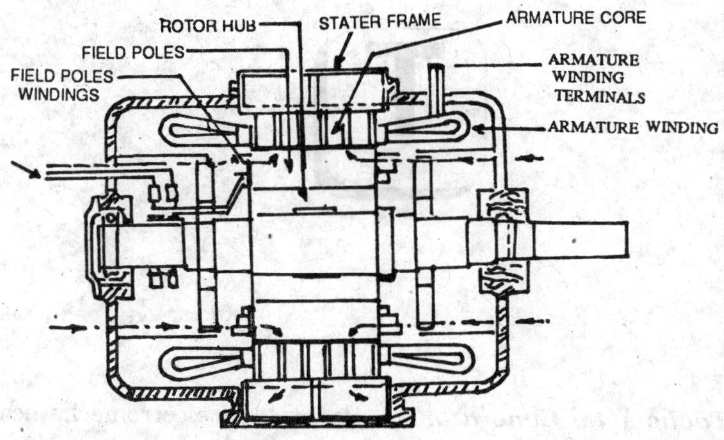

Fig. 19. Selectional view of synchronous motor showing main components.

Synthetic-resin-bonded Paper : Laminated sheet insula- ting material having papers bonded together under heat and pressure with synthetic resins. It is used for panels, terminal boards, coil flanges, etc.

System Function : Refers to a network function which describes the kind of response that will result from a given stimulus (*e.g.* applied voltage).

Tachometer Generator : Refers to an electromechanical device which is used in servomechanisms. It resembles a small motor and is having an output voltage proportional to its shaft speed. Units that produce a direct-voltage output usually have a permanent-magnet field excitation. A.C. units get excited by the alternating supply, and produce an output voltage of supply frequency, the polarity of which gets dependent on the direction of shaft rotation. It is also called a rate generator.

Tandem-compound Turbo-alternator : It is a turbo-alter- nator in which the turbine is having two or more cylinders in line driving a common shaft.

Tank Capacitor : Refers to an industrial capacitor in which the dielectric and electrodes have been laid up in a roll, these elements then being joined in series-parallel to suit the rated operating voltage and output. The tank allows adequate space for free thermal circulation of the oil, and its size has been determined by the heat to get dissipated from the capacitor losses, which have been normally 0.2-0.4% of the capacitor rating. Tank capacitors have been rated up to 500 kV At or above.

Tank Circuit : In an electronic oscillator, this term refers to the tuned LC circuit in which the stored energy is circulating at a frequency determined by the circuit parameters. The energy gets dissipated in tank-circuit losses, and may also be drawn upon for an output load (*e.g.* for high-frequency heating or for aerial radiation).

Tap Changing : The term used for the method of controlling the voltage ratio of a power transformer, by tapping the windings in such a way as to charge the number of primary or secondary turns. Tap-changing may carried out when the transformer has been out of circuit; or, particularly with short period and daily load cycles requiring voltage adjustment, when on load.

Tariff : In electricity supply, this term is used to refer to the price list of the supply authority who is detailing the charges made to the consumer in accordance with the characteristics of the supply.

Telecommunication : The remote communication of audible or visible information by means of electric signals in line or radio transmission. Telephony is basically telecommunication of sound, television that of vision, and telegrphy that of signs and characters like letters and numbers as well as of facsimilies. Teleprinting is a form of telegraphy; a keyboard transmitter.

Tension Insulator : An insulator, in an overhead-line system, which has been capable of transmitting the tension of the conductor to the supporting tower. It is also called a strain insulator.

Terminal : In an electric device, a component that is designed to connect the device to an external conductor.

Terminal Pair : Alternative term for port.

Terminal Tower : A lattice tower situated at the end of an overhead line the, and designed to withstand the longitudinal load of the phase conductors. It is also called a dead-end tower.

Tertiary Winding : An auxiliary winding,additional to the normal primary and secondary.

Tesla : It is the SI unit of magnetic flux density (symbol : T).

Tesla Coil : Induction coil which is used to develop a high-voltage discharge at a very high frequency. A high-voltage transformer (Fig. 1) promotes a discharge across the gap G_1 and charges capacitor C. Low-frequency high-

amperage currents are able to osciliate in winding P, inducing high-frequency high-voltage oscillations in S. These may produce dicharges across the gap G_2. It is also called a Tesla transformer.

Test Set : Refers to a multi-range instrument which is used for measuring two or more quantities, *e.g.*voltage, current and power. Both single and double instrument test sets are in use. For d.c test sets,moving-coil instrument are always incorporated. For a.c., moving-iron insturments are generally incorporated, but when power is to be measured in addition then an electrodynamic or induction wattmeter must be added; sometimes all three instruments have been of the same type.

Test Shield : Of a cable, this item refers to a metal sheath which is situated under the lead and insulated from it (cf. earth shield).

Testing Joint : Of a lightning protective system it is a joint in the system which is designed to facilitate the measurement of resistance.

Tetrode : An electronic device having four electrodes, normally cathode, anode, control grid and screen.

Thermal-agitation Noise : Refers to the unwanted electric phenomena in electric circuits, particularly those having amplifiers. They have been the consequence of the thermal agitation of free electrons in a conductor.

Thermal Breakdown : The term used for an electric breakdown resulting from a runaway condition that is caused in an insulating material by self-heating from its internal losses, or in a semiconductor junction by the production of free charge carriers resulting from excessive temperature rise of the junction. The breakdown may also occur in an insulating material by contaminants which are conducting and overheat to produce carbon that produces conducting tracks through the material.

Thermal Ohm : Refers to a measure of thermal resistance. It may be defined as the thermal resistance of a body which has a temperature difference of 1°C between oppposite faces when heat flows at the rate of 1 W.

Thermal Overload Relay : It is an overload relay that makes use of the heating effect of current for operation.

Thermal Power Station : Refers to a power station where electric energy gets produced by conversion from thermal energy obtained from fuel-fired plant.

Thermal Protection : Refers to the protection against damage by overheating. This may be achieved with an overload relay like a thermal overload relay, or a simple thermal relay.

Thermal Relay : It is relay that disconnects apparatus when the temperature exceeds the safe value. Such a device could work directly from an electric signal which is produced by a hot-bulb or resistance thermometer placed at a suitable point in the apparatus to be protected.

Thermionic Emission : Refedrs to the emission of electrons from a substance when heated. Emission occurs from a hot metal or semiconductor when its electrons have been given sufficient thermal (kinetic) energy to surmount the surface potential barrier.

Thermonic Rectifier : Refers to a thermonic valve or tube in which the undirectional electron flow is used for rectification.

Thermionic Relay : Thermionic valve, generally a thyratron which is operated so that the application of a comparatively small impulse to the control electrode allows the passage of a large current.

Thermistor : Refers to a resistor whose resistance alters substantially with temperature, a semiconductor material providing a very large negative temperature coefficent. The type of thermistor can be used in circumstances which need a resistance varying with power dissipation, or it may be used to measure small temperature changes.

Thermocouple : The term used for a device which uses the Seebeck effect to measure temperature. Of the available dissimilar metals, antimony/bismuth provide the highest output, but iron or copper/constant have been more durable and stable.

Thermocouple Generator : Refers to a form of thermo-electric converter. The normal thermocouple has been inefficient as a converter due to the that-condition loss between the hot and cold junctions. If one conductor has been placed by a plasma, or semiconductor materials are used, a good electrical conductivity could be combined with a poor thermal one, so increasing the thermal-electrical conversion efficiency.

Thermocouple Relay : Refers to a combination which consists of a thermocouple connected to operate a sensitive relay when some pre-arranged temperature is reached. The thermocouple may be associated with a heater carrying the control current so that operation could be obtained as for the thermal relay.

Thermoelectric Converter : A device which is used for the direct conversion of heat into electric energy. Various approaches include the fuel cell, thermo electron convertor, thermocouple generator, and magnetohydrodynamic generator.

Thermoelectric Effect : The term used for an interaction between an electric and a thermal condition which arises in a material or a combination of different materials.

Thermo-electron Converter : Refers to a form of thermo-electric converter. When electrons are emitted from a hot cathode, a calculable fraction of their number will have a velocity high enough to reach a nearby anbode at a small negative potential, in spite of the opposing potential barrier. The electrons lose their kinetic energy at the anode surface, and the potential energy that results could be used to drive a current through a load circuit which is between anode and cathode.

Thermometal : Alternative name for bimttal.

Thermopile : A device which is used for converting heat directly into electric energy. It generally consists of a battery of series-connected thermocouples.

Thermostat : An apparatus which is used for automatically controlling temperature, either at a predetermined valve or within specified limits, by adjusting the flow of heat into or out of the enclosure body, or fluid, whose temperature it has been controlling. A thermostat consists of :

(a) a temperature-sensitive element generating a mechanical force or electric signal, and

(b) a regulator which is either an electric switch or a fluid valve or some comparable device. The regulator directs or indirectly adjusts the flow of the heating or cooling medium in accordance with the response it receives from the directing element.

Thomson Effect : It is a thermoelectric effect. If different parts of the same metallic conductor are maintained at different temperatures, a temperature-gradient will exist and a heat flow will occur in the metal. If now an electric current is passed through the conductor, the temperature distribution will get disturbed. The accompanying evolution or absorption of heat (other than the Joulian heat) comprises the Thomson effect.

Three-ammeter Method : It is a method of measuring small single-phase powers. In Fig. 2 the load is having a phase-angle ϕ and takes I_0 amperes at V_0 volts. By using three instruments, and ignoring any of their impedance effects, the load power P_0 and power factor $\cos\phi$ could be obtained. From the phasor diagram, we know

$$I_2^2 = 1_0^2 + I_1^2 + \approx I_0 (V_o/R) \cos\phi$$

hence

$$P_0 = V_0 I_0 \cos\phi = \frac{1}{2}(I_2^2 - I_0^2 - I_1^2)R$$

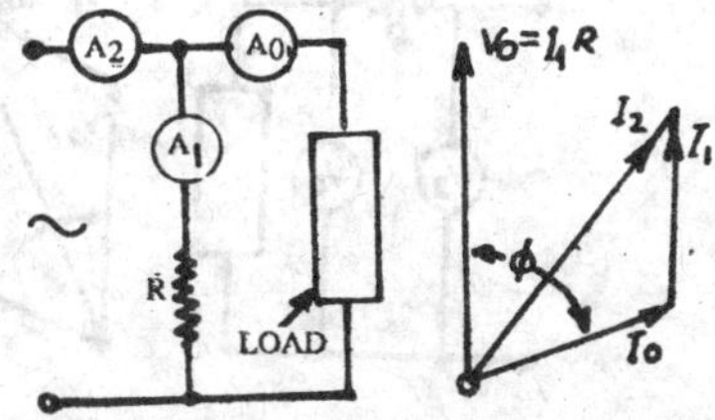

Fig.2 Three-ammeter method of measuring small single phase powers

and the power factor has been as follows :

$$\cos\phi = \frac{1}{2}\,(I_1^2 - 1_0^2 - I_1^2)I_0I_1$$

Three-phase Four-wire System : Refers to a system of distribution which is using four conductors, three of which get connected to the three-phase supply and the fourth having its termination on a neutral point in the source of supply.

Three-phase System : Refers to a polyphase system in which the number of phase M has been 3, and the phase displacement has been 2π/3. It has been the system of distribution, transmission and generation in almost universal use for electric energy in bulk, the phase voltages being as far as possible balanced.

Three-voltmeter Method : It is a method which is used for measuring small-phase powers, similar to the three-ammeter method. With the same symbols as in the latter case (Fig. 3),

$$V_2^2 = V_0^2 + \frac{1}{2} + V_4^2\, 2\, V_0 I_0\, R \cos\phi$$

giving

$$P_0 = V_0 I_0 \cos\phi = ½\,(V_2^2 - Vsub0^2 - V_1^2)R$$

and

$\cos\phi = 1/2\,(V_2^2 - V_0^2 - V_1^2)V_0V_1$ The three-meter methods owe the disadvantages that, by requiring the

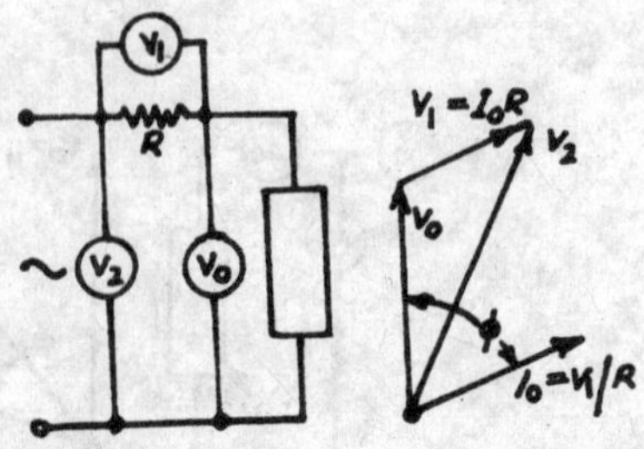

Fig. 3 Three voltmeter method of measuring small single phase powers.

difference between squared values, the errors in the instruments would get magnified.

Three-wire System : Refers to a distribution system, which is used for direct current or single-phase alternating current, and is using two conductors and a neutral wire. The supply has been taken between one conductor and the neutral, the latter being usually earthed and carrying only the difference current.

Thrustor : Refers to an electrohydraulic mechanism which is able to convert electric energy into a straight line, constant thrust. A centrifugal pump at the bottom of an oil-filled tank would force a piston upwards wih an constant force. At is a control device, and could be used on a mechanical brake.

Thury System : A system of high-voltage d.c. transmission in which several series would generators are connected in series, the loads usually being motors driving auxiliary generators in substations. The voltage of the main generators gets varied to maintain the current constant.

Thyration : Refers to a gas-filled triode or tetrode which is now being replaced for many applications by the thyristor.

Thyristor : Refers to the generic name for a group of bi-stable semiconductor devices which have three or more junctions, for example silicon controlled rectifier (an alternative term for reverse-blocking triode thyristor), diac (the alternative term for bi-directional diode thyristor) and triac (the alternative term for bi-directional diode thyristor). Thyristors have been suited in arrangements to control the supply of power between a source and the load.

Tidal Power : The term used for the power resources which are available due to the rhythmical movements of the seas. A few practical schemes are in existence for converting this power into electric power.

Tie Line :

(1) Refers to an interconnecting line in a distribution system.

(2) Also, refers to a link between two or more private automatic telephone exchanges.

Time Constant : The time in which a function like the charge on a capacitor or current in an indicator, falls to I/e of its initial value (I/e≈ 0.368). If C denotes the capacitance, L the inductance, and R the total series resistance, the time constants of capacitive and inductive circuits would be R C and L/R respectively.

Time-control Acceleration, Starting : A method which is used for automatical controlling the acceleration of a resistance-started electric motor during the starting period. Use has been made of one or more accelerating relays of the time-delay type for determining the period that elapses between the closing of successive accelerating conductors. The acceleration period of the motor has been independent of the load conditions.

Time/Current-control Starting : Refers to a method of automatically controlling the starting of a resistance-started motor. It has been similar to time-control starting above, but incorporates a timing relay designed to get influenced by the magnitude of the motor current during acceleration in such a manner that the time delay increase gets with motor load. Thus, a motor lightly loaded would possess the accelerating contractors closed more rapidly than if it were heavily loaded.

Time of Recovery : In an automatically regulated circuit, this term is used for the time needed after a specified disturbance for the voltage or current to recover to 95% of its original value.

Tip Speed Ratio : Refers to an important design figure for wind-power generating plant. It has been the circumferential speed of the rotor blade tips divided by the wind speed and usually lying between 5 and 10 for such plant.

Tirrell Regulator : Refers to an automatic voltage regulator of the vibrating contact type. A regulating field rheostat rapidly gets switched in and out of circuit to correct deviations from the normal value of voltage.

T-Joint : Refers to a form of branch joint in which the branch cable has been taken away at right-angles to the main cable. It gets inclosed in a pressed lead or copper box of suitable shape.

T-network : A network gets three impedances connected to gether at one end. One free end gets connected to an input terminal, another to an output terminal, and the third free end to both second input and the second output terminals, as shown in Fig. 4. It is also called a star or Y-network.

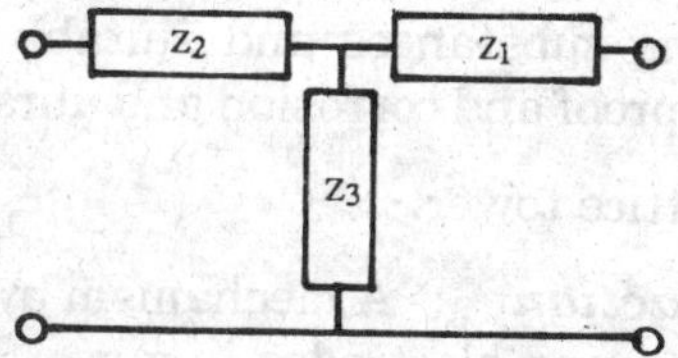

Fig. 4. Symmetrical unbalanced T-network

Tong Tester : A portable test equipment which is used for measuring current. It consists of a secondary winding which is wound upon one leg of a rectangular frame core and generally connected to an integral ammeter. One corner of the core is having a butt joint, and one side adjacent to the butt joint is hinged so that the core can be open and closed around a cable or busbar carrying the current to be measured.

Toroidal Winding : Alternative name fore ring winding:

Torque-amplifier Selsyn : A transmitter selsyn which is possessing a special arrangement of windings and involving both a commutator and sliprings, whereby the torque reaction at the transmitter has been largely neutralised. The unit is having both the functions of the torque amplifier and transmitter selsyn, requiring much less manual effort.

Torque Angle : Alternative name for load angle.

Torque Motor : Refers to a motor which makes a part of a revolution when current gets applied, to exert a torque against some controlling force such as a spring. It has been used for the operation of breakes, rheostats, etc.

Total Break Time : Of a circuit-breaker, this term refers to the time elapsing between the application of tripping power and the extinction of the arc cause by the opening of the contacts.

Totally Enclosed Machine : Machine in which there occurs no access whatever to the inside of the casing, so that cooling must gets effect by conduction to and subsequent dissipation from the external surface.

Tough-rubber Sheath : A protective covering which is applied to an insulated cable. It consists of a rubber mixed with hardening substances and suitably vulcanised to make it waterproof and corrosion and abrasion resistant.

Tower : See lattice tower.

Townsend Conduction : A mechanism by which a spark takes place between electrodes in a gas. It is dependent on the production of an electron at a point near the cathode, to precipitate an electron avalanche.

Tracking : Refers to the formation, along the surface of an insulating material, of carbon conducting paths which are caused by surface Stresses. Fibrous material such as paper, cotton or asbestos have been generally very susceptible to tracking.

Traction Battery : Battery for use in the power unit of battery-driven vehicle.

Traction Motor : A motor which has been designed for use in electric traction. A series-characteristic motor may be ideal for this work.

Train Line : A cable which is extending the length of railway coach and terminating in sockets at each end. Couplers between each coach make electric continuity to be maintained through the length of the train.

Transducer : Refers to a device which, under the influence of a change in energy level of one form or in one system, gives rise to a corresponding change in energy level of another form or in another system. The ideal correspondance between the input and output is usually regarded to be an output which is proportional to the input, or to its derivative or integral. In such a case the device has been linear transducer. In a reversible

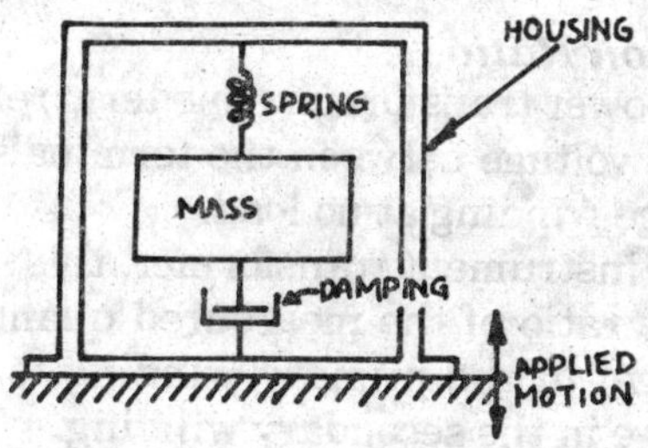

Fig. 5. A seismic transductor

transducer the action has been reversible *i.e.*, the correspondence may be in either direction. See Fig. 5.

Transductor : Refers to an electromagnetic device having one or more ferromagnetic cores and windings designed to use saturation effects in the magnetic circuit in a manner to vary an alternating current or voltage by an independent current, usually a direct current.

Transfer Admittance : Refers to the ratio of the current flowing in any part of a network to the electromotive force applied elsewhere in the network that produced that current.

Transfer Function : Refers to a type of system function. It relates the response at a given pair of terminals to source or stimulus at another pair of terminals. Its usual form has been K.G.(p), where K is a multiplier or 'steady-state gain', and G(p) is a function of time in terms of p=d/dt and the parameters of the network.

Transfer Impedance : Refers to the ratio of the voltage applied across two given terminals in the network to the current flowing at some other point in the network.

Transfer instrument : An instrument which has been so constructed that it can be calibrated on direct current and used to indicate a.c. quantities without sensible error. An example has been an electrodynamic instrument.

Transfer Switch : A switched which is used for transferring load current from one tap-selector switch on a transformer to another, so that each is not carrying a current when it operates.

Transfer Tripping : See intertripping.

Transformation Ratio

1. Of a power transformer, this term refers to the ratio of the voltage between the terminals of the higher-voltage winding at no load.
2. Of an instrument transformer, this term also refers to the ratio of the mearsured quantity (current or voltage) in the primary winding to the current or voltage in the secondary winding.

Transformer : Refers to a static electromagnetic device which consists of two electric windings or circuits (Fig. 6) magnetically interlinked by an iron core, whereby an alternating voltage applied to one of the windings produces, by elecrtromagnetic induction, a corresponding voltage in the other winding. The winding to which the voltage is applied is termed as the primary winding, the other being the secondary winding. Transformers are used for converting energy received at one voltage to energy transmitted at a different voltage; for separating electrical circuits while allowing power flow between them and for matching impedances in communication circuits so that maximum power could obtained from a source.

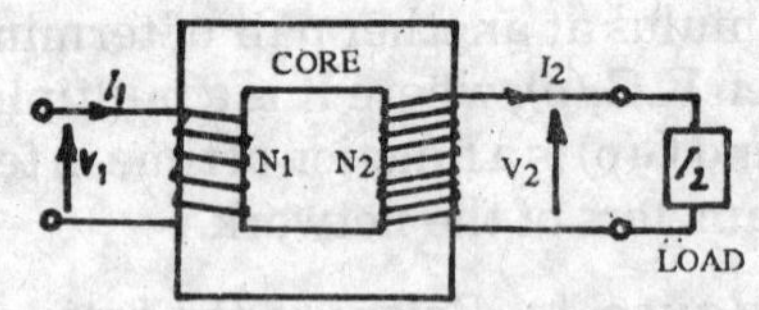

$$\frac{V_1}{V_2} = \frac{N_1}{N_2} = \frac{I_1}{I_2}$$

Fig.6 Elementary transformer

Transformer Hum : See noise.

Transformer Kiosk : The term used for a weatherproof enclosure for a power transformer and its associated swtichgear, having no additional space.

Transformer Oil : See insulating oil,

Transformer Ratio : See transformation ratio.

Transformer Substation : Refers to a substation which is installed at a point on a network where a change of voltage has to be carried out. It has been therefore to be found on sites adjacent to generation stations where power is received at generation voltage and gets stepped up to the grid voltage. These stations have been also important switching points where feeders from a number of directions could be marshalled through circuit-breakers on to the common busbars.

Transformer Tank : Refers to a container in which a transformer gets immersed in a cooling and insulating medium. It may carry fins for providing a heat-dissipating surface for a transformer with natural oil cooling.

Transformer Voltage : A voltage which is induced by transformer action in coil or other conductor that gets linked with an alternating flux.

Transformeter : Combination of an integrating meter and associated current transformers in a single unit. It has been available for a single-phase or polyphase circuit.

Transient : The term used for the variation of a system quantity (such as current, flux, velocity) in a physcial system following a disturbance. The variation persists, normally until the system gets settled down to a new energy state.

Transient Reactance : Reactance of the armature winding of a synchronous machine which is caused by the leakage flux.

Transient Stability : Stability which follows sudden disturbances, such as sudden load increases on a power system, or switching operations or faults.

Transistor : The solid-state counterpart of the thermionic valve.

Transistor, Tetrode : A transistor in which arrangement has been made to apply a transverse voltage across the base layer, so that the emitter current gets restricted to a smaller area, decreasing collector capacitance and increasing maximum frequency.

Transition Inductor, Resistor : An inductor or resistor which is used in on-load tap changing equipment.

Transmission Dynamometer : Refers to a device which is used for measuring the mechanical output of a prime mover or electric motor transmitted to another equipment. Simple mechanical transmission dynamometers consists of an epicyclic gear or belt in the transmitted drive.

An electric dynamometer consists of a d.c. generator that converts the output of the machine under investigation to electric power, this being then dissipated in resistors, or delivered to a suitable power network. The field system of the generator get mounted on trunnions, and the torque which are used determined from the force needed to keep it stationary.

Transmission Line :

(1) Refers to a system of conductors which are used for the overhead transmission of electric energy from a generating station or substation to other stations.

(2) Refers to a theoretical equivalent circuit which is representing a power or telephone overhead or cable line.

Transposition : The term used for the variation of the relative position of parallel conductors to reduce or

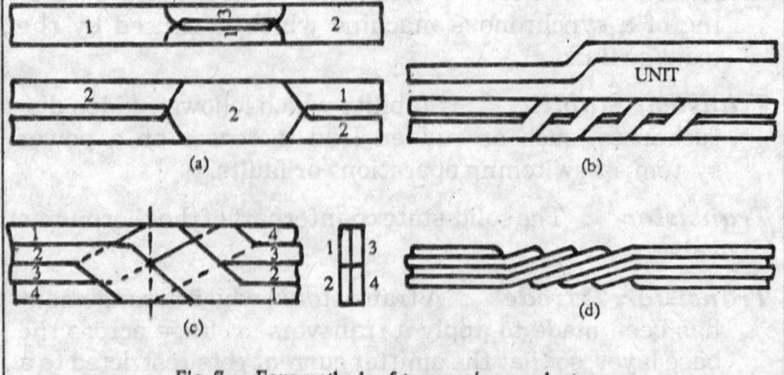

Fig. 7. Four methods of transposing conductors.

eliminate unwanted inductive effects. In the case of power lines, the harm occurs due to the resulting asymmetry in line voltage drops by which the terminal voltage could be affected, possibly resulting an unbalance. To avoid this, the three (or six) conductors of an overhead circuit get transposed in the course of the line run so that each phase conductor becomes in turn the upper conductor. For overhead telephone lines, if the relative positions of the pair gets unchanged there will occur interference and crosstalk. Transposing has been simply carried out by taking each two pairs and rotating them cyclically so that the phase of the induced voltage is changed by 180° at each stage. Four methods of transposition are shown in Fig. 7.

Travelling Cable : On an electric lift car it refers to a flexible cable providing electric connection between the car and a fixed point.

Travelling-wave accelerator : A linear acelerator which is used for the production of high-energy electrons.

Trembling Bell : Bell having a mechanism giving a trembling action. It is the most widely used type for alarm and general purposes.

Triac : A five-layer semiconductor device which is similar to a diac but with a gate electrode.

Triboelectricity : Electric charge which is produced by friction. Substances may be arranged in a triboelectric series so that any substance will attain a positive charge when rubbed with another substance below it in the series, or a negative charge when rubbed with a substance above it.

Trickle Charge : Refers to a steady, low-current charge which has been designed to maintain a battery in a fully charged condition.

Trifurcating Box : Refers to a closed box in which the cores of a three-core cable can be connected to external conductors.

Triode : A three-electrode electronic tube having a cathode, an anode and a control grid, or a similar semiconductor device. A gas-filled triode is called a thyratron.

Triple-n Harmonic : A harmonic of order that has been a multiple of 3, usually an odd multiple (*e.g.* 3rd, 6th, 15th.....). Such harmonics have been of importance in three phase systems and machines.

Tripping Fuse : A fuse having an auxiliary device that will trip other apparatus, like a switch.

Trolley : Refers to the fitting for an electrically-driven vehicle so that contact can be provided with an overhead wire by means of a trolley wire.

Trolley Frog : A device which finds used in overhead supply for electric traction. it gets incorporated at the junction of two wires to allow the passage of a current-collector along either wire.

Tropical Switch : A switch which is mounted on feet or bosses to ensure an air space between its base and the mounting surface. It safeguards the switch against the effects of excessively damp climates. It is also called a feet switch.

Troughing : A reformed channel which has been designed to accomodate cables and protect them from mechanical damage.

Trousers Joint : See breeches joint.

Truck-type Switchgear : Refers to a switchgear in which the circuit-breaker gets mounted on a wheeled truck together with ancillary items like current and voltage transformers, relay instruments and control switches. The truck rolls on the floor and gets guided into a housing containing busbars and circuit connections which are selected by means of plugs.

Trunk Feeder : A feeder which connects two sources of electric energy, or two networks. It is also called a trunk main or interconnecting feeder.

Trunking : Cable casing which is constructed of metal, wood or insulating material, and usually of rectangular cross-section. One side has been normally removable or hinged for the whole of its length to allow cables to be inserted.

Tube-sinking : Refers to a method of applying an aluminium sheath to a cable. It involves pulling a cable core into a long aluminium tube, and then pulling both through a sinking die so that the sheath has been brought close to the core. The cold working makes the metal harder and less ductile. Tubes sufficient to cover a 250 m length of cable could be produced.

Tumbler Switch : Refers to a small single-pole switch having a quick-break action operated by a lever handle pivoted near the face of the switch. It finds use in low-power circuits such as those controlling individual lamps.

Tuned Circuit : Network consisting of an inductive and a capacitive element, one or both of which can get varied to alter the resonant frequency.

Tungsten : A metal with a very high melting point which is used widely as a lamp filament. It is also employed in some thermiontic valves and for sparking points and contacts. Tungsten steels were used for permanent magnets.

Tungsten Arc

(1) Refers to an arc between tungsten electrodes, the radiation being principally from the incandescent electrodes.

(2) Also, refers to an arc through tungsten vapour, having a characteristic emission.

Tunnel Diode : Refers to an heavily doped semiconductor diode that exhibits negative resistance over part of its I/V characteristic. It finds use in oscillators and low noise amplifiers.

Turbine : See steam turbine, water turbine.

Turbo-alternator : Refers to an alternator which is designed for high-speed operation, coupled to steam or gas turbines. It has few poles and forms the most important class of machine for power generation in the world today.

Tubovisory Equipment : Refers to the steam-turbine supervisory equipment. Modern turbines need special instrumentation, particularly for starting. This is arranged to indicate and/or record the following parameters :(a) shaft eccentricity: (b) differential expension between rotating and stationary parts; (c) overall expansion of the turbine; (d) vibration of bearing pedestals.

Turns Ratio : Refers to the ratio of the number of turns in phase winding of a transformer that gets associated with the higher voltage to the corresponding number of turns in the lower-voltage winding.

Two-fluid Cell : A voltaic cell in which the anode and cathode get immersed in different electrolytes.

Two-in-hand winding : Refers to a method of winding a lap-connected armature in which two circuits have been wound simultaneously.

Two port Network : Refers to a passive electric network having one input and one output pair of terminals, in the past also called quadripole or four terminal network.

Two-reaction Theory : Refers to an analysis of the behaviour of a rotating electric machine by regarding the magnetomotive forces acting along the direct axis and the quadrature axis. The former has been the direction in which the field winding magnetises, the latter has been electrically in space quadrature to it.

Two-terminal Network : See two-part network.

Two-way Switch : Alternative name for a landing switch.

Type Test : A test which has been taken on a device or equipment when first produced, to determine whether its design meets the specification.

U-link : A short length of rigid conductor in the shape of a U which is used to make connection between appropriately spaced sockets or plugs mounted in a plain surface.

Ultrasonic Flow Detection : Refers to a method of detecting flaws in a specimen which is based upon impulse reflection. A crystal probe is kept upon the specimen, usually with a coupling medium such as oil between. Ultrasonic impulses enter the specimen from the point contact in a closely knit beam and get reflected back by flaws such as cavities, cracks or inclusions. These get reproduced on a cathode-ray screen and compared with the wave reflected from the back of the specimen.

Ultrasonic Wachining : Methods of removing metal,by using spark machining or impact grinding. Ultrasonic drilling is applied to the production of dies in tungsten carbide and hardened steel).

Ultrasonics : Refers to the study and application of acoustic wave of frequencies above the range of human hearing. It is also called supersonics.

Ultra-violet Radiation : Electromagnetic radiation of frequencies which is higher than those of visible light, and wavelengths of 0.39 - 0.005 μm.

Unbalanced, Unsymmetrical There Phase System : A three conductor three-phase system is said to to be symmetrical when the phasor diagram of the three line currents I_1, I_2, I_3 has been an equilateral triangle, see Fig. 1(a), and has been unsymmetrical when the diagram is an ordinary triangle such as in Fig. 1(b). A four conductor three-phase system has been unbalanced when the phasor diagram of I_1, I_2, I_3 together with the neutral-conductor current I_N has been a quadrangle, see Fig. 1(c), when L_N=O the system has been either unsymmetrical or symmetrical according to Fig. 1(b) or (a) respectively.

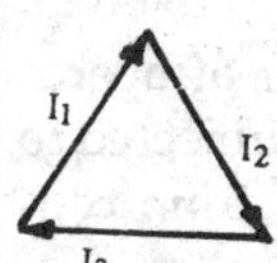

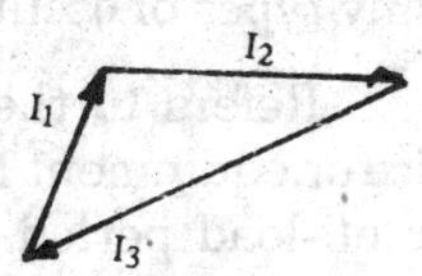

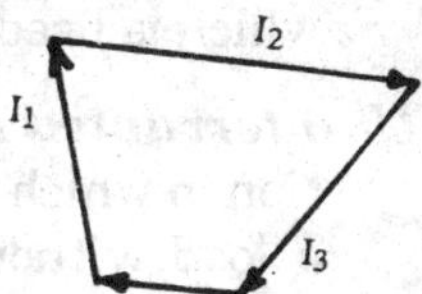

Fig. 1. Current phasor diagram in a three-phase system, (a) symmetrical system, (b) unsymmetrical system, (c) unbalanced system.

Undercompensation : Refers to the result of a compensating winding which is having less effect than the armature reaction it opposes. The output votage will tend to decrease as the load current increases.

Underground Distribution : The term used for the system of distribution in general practice in urban and sub-urban areas wher overhead systems have been likely to be dangerous or unsightly or to cause too much obstruction. Both high-voltage and low-voltage systems could be laid underground, using insulated cable of suitable type and laid with necessary safeguards.

Undervoltage Release : A device that opens a switch when the supply to a circuit falls below a given voltage. A trip-coil gets incorporated for this purpose in such equipment as hand-operated motor starters so that the motor does not re-start when an interrupted supply is restored. It is also called a low-volt release.

Unearthed System : A system having no conductive connection to earth. It has the advantage that should be an earth fault occur on any line conductor there occurs little interference to the system operation on load.

Unexcited Synchronous Motor : A single or three-phase synchronous motor which does not have no form of d.c. excitation.

Unijunction Transistor : A two-layer semiconductor device which has a p-type rod placed asymmetrically in an n-type base. There are two contacts to the n-region, base 1 being nearer to the emitter than base two. The device will conduct only at certain value of emitter-base one voltage (the value of which can be varied by the biasing arrangements). It is therefore a triggerable device which is used in many types of oscillator circuit.

Uninterrupted Duty : Refers to the condition of operation in which a device or equipment has been subjected to a load without any off-load period. Different from continuous duty.

Unipolar Machine : See homopolar machine.

Unipolar Transistor : Refers to a transistor in which conduction of current is dependent only on one polacity of charge carriers; for example a field-effect transistor.

Unit Capacitor : Refers to an industrial capacitor in which the impregnated paper dielectric and electrodes get encased, usually in rectangular containers, so that the elements substantially will fill the whole of the space in the main container, providing good terminal contact between the elements and the container to aid in dissipating the heat from the dielectric. Connection between the capacitor terminals and the electrode foils of the elements has been done by soldering connections either directly to the projecting edges of wide electrode foils, or to metal tabs inserted in the element winding. Unit capacitor ratings have been lower than those of tank capacitors.

Unit Magnetic Pole : Refers to a magnetic pole such that, when it hs been one centimetre from a similar pole in a vacuum, each exerts a force of one dyne on the other.

Unit-step Function : Refers to a function of time, such that its value has been zero for all values of t up to t=O and has been unity for all t from t=O to t+α. Its principal application in transient analysis is as a switching function. Thus, a lightning surge may be idealised into unit-step function shape; and closing the switch separating a d.c. source from a circuit has been equivalent analytically to the application of a unitfunction voltage to the circuit.

Unit Testing : Refers to a form of direct testing, which is applied to a unit, or group of units, of a multi-unit circuit-breaker. By this means, it becomes possible to assess the behaviour of the complete circuit-breaker up to its rated making and breaking capacity from tests made one or more of its units.

Uninterrupted Duty : Refers to the condition of operation in which a device or equipment has been subjected to a load without any off-load period. Different from continuous duty.

Universal Bridge : A commonly used laboratory bridge (usually self-contained) which is using three built-in-arms, *viz.*, two resistors P and Q, and a parallel capacitor-resistor combination (Fig. 2). It could be switched to a real ratio form for measuring capacitance and to a real-product form for the measurement of inductive unknowns (see a.c. bridge), in each case giving also an inducation of

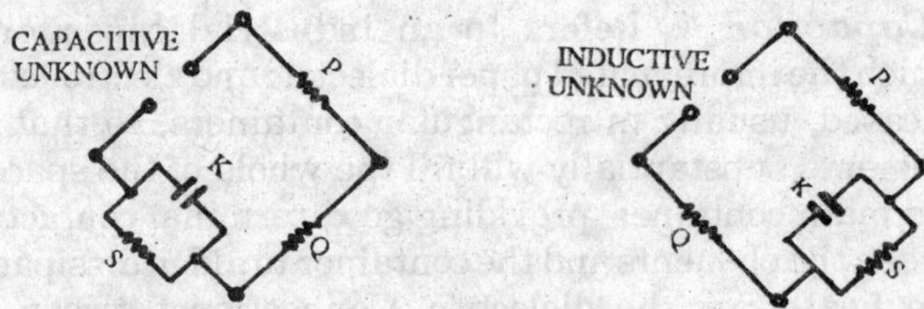

Fig. 2. A universal bridge

power factor from the balance relation of the capacitance and resistance arms P and Q have been adjustable so that the range could be extended up and down. Taking the balance conditions respectively for (*i*) a parallel CR and (*ii*) a series LR unknown, the balance conditions would be

(*i*) $$\frac{R}{I + j\omega CR} = \frac{P}{Q}\frac{S}{1 + j\omega KS}$$

whence

$$R = S(P/Q) \text{ and } C=K (Q/P)$$

(*ii*) $$R + j\omega L = PS[1/Q) + j\omega K]$$

whence

$$R = S(P/Q) \text{ and } L=K(P/S)$$

Both balances have been independent of frequency.

Universal Motor : A commutator machine which has been designed to work on d.c. and single-phase a.c. supplies without alteration. It normally has a circuit identical to that of a d.c. series motor, and its construction has been the same except that it becomes necessary to laminate the field structure as well as the armature core.

Universal Test Set : A test set which is used particular for radio and other low-current circuits. It measures direct and alternating quantities and resistance. Such instruments have been using a rectifier for alternating current, with provision for excluding the rectifier when direct current is measured.

Uasymmetrical Three-phase System : See unbalanced, unsymmetrical three-phase system.

Upset Butt Welding : Refers to a variety of butt welding in which pressure has been first applied to the joint and then current passed until welding temperature is reached and the weld is 'upset'. It is also called slow butt welding.

Urea Resin : Moulding (thermosetting) material which is produced by the controlled reaction of urea and formaldehyde. Its characterisitcs are similar to those of phenolic resin. It is used for the production of plugs, sockets, ceiling roses, housing, switch-plates, etc.

U.V. : Ultraviolet light.

Vacuum : A space from which gas is removed. A perfect vacuum, in which the gas pressure has been zero, is unattainable, but a diffusion pump has been capable of producing down to 10^{-7} mmHg. Vacua of the order of 10^{-6} mmHg are used in high-vacuum valves and other electron tubes. Vacuum pressures are generally measured in conventionl millimetres of mercury or in torrs. 1 mmHg=1.0000001 torr. 1 torr=1/760 atm. (1 mmHg=133.32 N/m^2).

Vacuum Impregnation : Refers to the filling of voids in a dielectric material by reduction of pressure, to increase breakdown strength and permittivity. Mineral oil, paraffin wax, petroleum jelly, etc., could be used to fill voids.

Vacuum Switchgear : Refers to a mechanical switch or circuit-breaker whose contact pieces operate in a vacuum.

Valley Value : Refers to the least instantaneous value of a time-dependent quantity within a given time interval which has been equal to a period for a perodic quantity.

Valve : This term was first applied to a device which allowed a current to pass through it in cone direction only; later this term was extended to cover developments of the original valve whereby it was enabled to amplify signal voltages and to generate electrical oscillations. The envelope of an electron valve could be either evacuated or gas-filled. High-vacuum valves are having thermionic cathodes; gas-filled valves may have thermionic or cold cathodes. Valves could be classified by the number of electrodes they contain as diode. (2) triode (3), tet-rode (4), pentode (5), hexode (6), heptode (7), octode (8), nonode (9), etc. In modern terminology the term electron tube is frequently used instead of valve.

Valve Rectifier : See thermionic rectifier.

Valve Voltmeter : Refers to a voltage-measuring device in which the alternating voltage to be measured has been rectified by a thermionic valve and the resultant direct voltage measured by a conventional d.c. instrument. Its input impedance can be made very high so that the instrument puts no appreciable load on the source to be measured, and it can be employed to measure alternating voltages of any frequency with equal accuracy.

Van de Graaff Generator : An electrostatic generator which is capable of providing a direct potential of up to about 8 MV of either polarity. It can be used to accelerate electrons or ions down an evacuated accelerator tube. The mechanical work expended in turning an endless belt against electrostatic force gets converted into electric-field energy (Fig. 1). Charges gets sprayed on to the belt from 'spray-set' needle-points at about 50 kV. The charge gets carried upwards to the interior of the h.v. electrode, a metal sphere to which it gets transferred by means of a second spray-set. Operation occurs inside a tank filled with pressurised insulating gas, *e.g.*, nitrogen-freon mixture at 1700 kN/m^2.

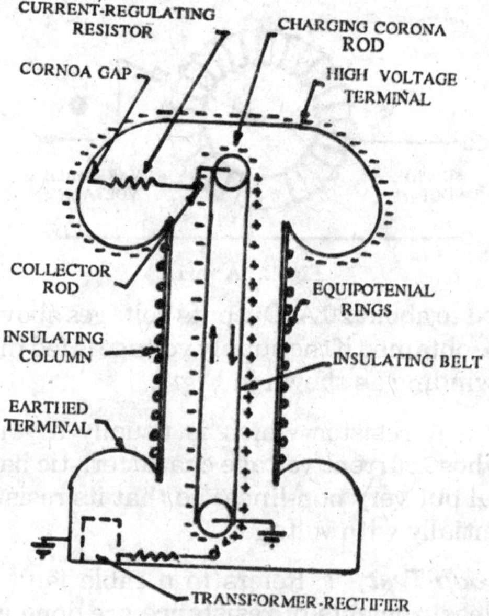

Fig. 1. Principle of a Van de Graaff generator. Electrostatic Charge is carried on the endless belt of insulating material to accumulate on an insulated sphere.

Var : Refers to the unit of reactive power.

Variable-capacitance Diode : A semiconductor diode which has been so designed that the capacitance between its terminals varies as a function of the voltage between the terminals.

Variable-speed Motor : A motor, the speed of which could be varied while remaining independent of the load.

Variable-voltage Control : Control scheme which provides a variable-voltage d.c. supply for the armature of a shunt motor.

Variac : A name applied to an auto-transformer which gets wound with enamelled wire on an iron core, generally toroidal in shape. A sliding contact is allowed to move on a bared track over the turns so that the voltage changes in steps corresponding to the volt-drop of a single turn. A double-wound arrangement having a bare secondary winding can also be employed. Losses are small, but the sliding contact (usually a carbon brush) limits the current

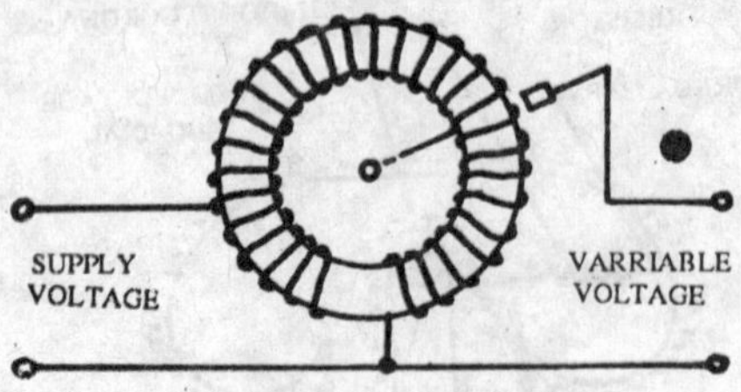

Fig. 2. A variac

collected to about 20 A. Outputs voltages above the supply could be obtained if the supply voltage is fed into a fraction of the winding as shown in Fig. 2.

Varistor : A resistor which is usually a semiconductor type, whose current/voltage characteristic has been symmetrical but very non-linear so that its resistance varies substantially with voltage.

Varley's Loop Test : Refers to a cable fault localisation test. Measurements of resistance are done with a resistance bridge, first with the fault forming one junction of the bridge, and second with the cable connector resistance measured directly.

Varmeter : An instrument which is used for the measurement of the reactive component of the volt-amperes in a circuit, and calibrated in reactive volt-amperes.

Vectormeter : An instrument which is having wo pointers. One is moving horizontally in proportion to power load, the other vertically in dependence on the wattless load. The intersection of the two pointers gives the load on a machine.

Ventilation : Of an electric machine, this term refers to the provision of access for cooling air while preventing the entry of dirt and moisture.

Vertical Plugging : Refers to the method of switchgear busbar selection in which the circuit-breaker has been mounted on a carriage arranged to move both horizontally and vertically. The sockets for duplicate busbars and circuit connections have been arranged in a horizontal plan. The required busbar has been plugging the circuit-breaker vertically upwards into the appropriate sockets.

V.h.f. : Very high frequency.

Vibrating-reed Electrometer : Refers to a vibrator unit in which a small capacitor is having its capacitance modulated by the steady vibration of one of its electrodes.

Vibration Damper : A device which is attached to overhead-line conductor to prevent excessive vibration being caused by the wind, etc.

Vibration Galvanometer : An instrument which resembles a direct-current-operated galvanometer, but producing an alternating torque. The reflected light from a small mirror on the moving part traces out a band of light, which gets reduced to a stationary spot when the current gets reduced to zero (as in the balance condition of an a.c. bridge network). The inertia of the movement restricts the width of the light band, and thus the sensitivity, unless mechanical resonance gets used to increase the response. The vibration galvanometer has been a useful bridge detector for industrial frequencies, and low audio frequencies up to about 250 Hz.

Vibrator : A device which is used for generating a mechanical vibration, in particular by electric means. Fig. 3 shows the essential features on an electromagnetic vibrator. When the coil, suspended in the annular gap of the permanent magnet has been supplied with alternating current of a suitable frequency, the force developed in

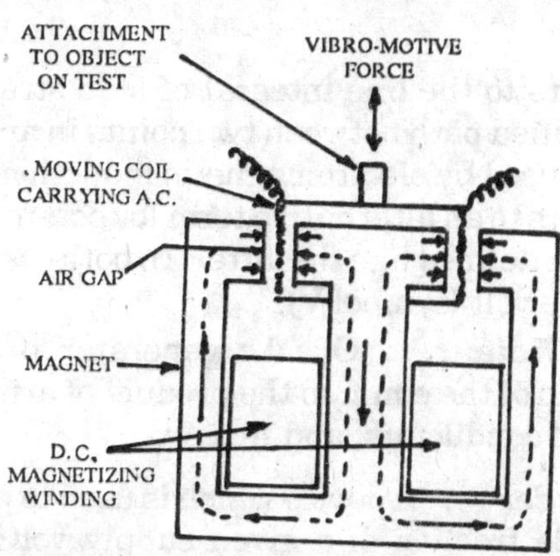

Fig. 3. An electromagnetic vibrator.

it gets transferred to a shaft that provides linear vibrational movement.

Vibrator Power Pack : A device which has been primarily intended to convert direct current at low voltage into direct current at high voltage without rotating plant. The basic feature has been a vibrating reed with appropriate contacts.

Villari Effect : Magnetostrictive effect which is appertaining to changes in the magnetisation of a specimen when it is mechanically stressed longitudinally or transversely.

Virtual Value : See effective value.

Volt : The SI unit of voltage and potential differences (Symbol:V).

Volt-ampere : Refers to the apparent power VI given by the product of voltage V and current I in r.m.s. values (Symbol VA). It has been normal to rate a.c. electrical apparatus in VA (or kVA or MVA) either as a matter of convenience (as with switchgear or rectifiers), or because such a quality better describes the service value of the device.

Volta Effect : Refers to the phenomenon which takes place when two dissimilar metals are kept in mutual contact in air, one acquiring a positive potential with respect of the other.

Voltage :

1. Refers to the line integral of field strength along a specified path between two points in an electric field produced by electromagnetic induction.
2. Refers to an alternative term for potential difference when defined like the latter. In both cases the SI unit is the volt (symbol V).

Voltage Coefficient : Of a d.c. generator, this term refers to the ratio of the e.m.f. to the product of armature speed, number of conductors, and flux.

Voltage Divider : A device which is used to permit a fixed or variable fraction of a given supply voltage to be obtained; often although ambiguously, termed as a potentiometer. Two forms are generally available, the most common being a resistor, across the ends of which the supply or reference voltage is applied.

Voltage Doubler : Refers to an arrangement of rectifiers which rectifies separately each half cycle of an applied alternating voltage, and adds the two resulting rectified voltages to give rise to a direct voltage that is having an amplitude approximately double that of the peak amplitude of the applied voltage.

Voltage Gradient : A term which is used interchangeably with electric field strength. It is measured in volts per unit length within an electric field region, such as an insulant subjected to voltage stress. Strictly, the voltage gradient has been a vector with direction opposite to that of the electric field.

Voltage Grading : Refers to a measure applied to an insulator or insulation for reducing any substantial inequalities of voltage gradient along its length or thickness.

Voltage-regulating Relay : Relay which is used for automatic control of on-load tap changing. When the voltage varies the relay initiates a tapping change.

Voltage Regulation :

1. Refers to the variation of the terminal voltage of a generator or the secondary circuit of a transformer.
2. Also, refers to the maintenance of the voltage of a medium voltage (m.v.) or low voltage distribution system within ±6% of the declared voltage.

Voltage Regulator : Refers to a device that, supplied at a given input voltage, will provide an adjustable output voltage. Common types of regulator have been the resistance divider (Fig. 4), autotransformer and moving- coil

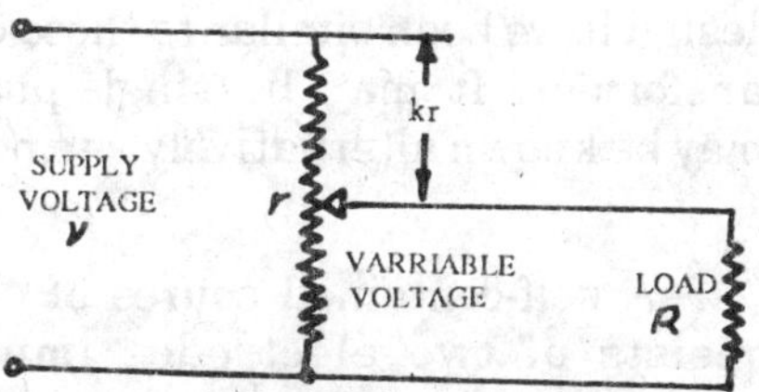

Fig. 4. A resistance-divider voltage regulator. For smooth voltage variation, r should never exceed 3R.

tance divider (Fig. 4), autotransformer and moving- coil regulator. A transformer with tap-changing may be employed for the same purpose, but then the output voltage gets varied in steps rather broader than those obtainable from other voltage-regulating devices.

Voltage Regulator Diode, Voltage Reference Diode : Refers to a semiconductor diode which has been designed to operate with zener breakdown and connected in a circuit analogous to Fig. 5. The voltage across the terminals of a regulator diode has been substantially constant throughout a specified current range and that of a reference diode is having a specified accuracy when the current is kept within a specified range.

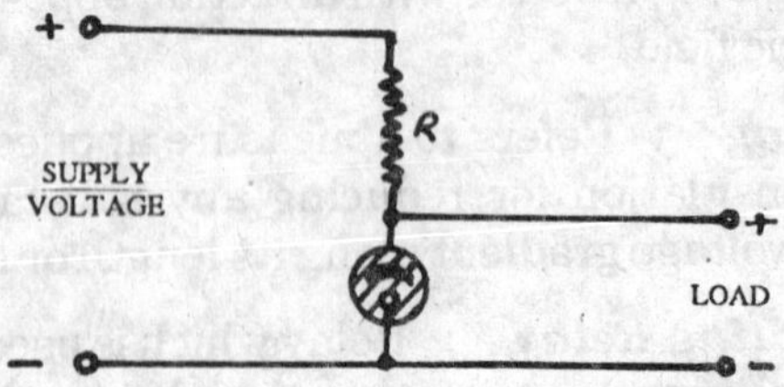

Fig. 5. Voltage stabiliser employing a cold-cathode glow-conduction tube.

Voltage Stabiliser : A device which is incorporated in a circuit to maintain a constant output voltage from a poorly regulated power supply and/or a varying output load. The simplest form of voltage stabiliser is making use of the substantially constant voltage characteristic of a cold-cathode glow-conduction tube when connected as shown in Fig. 5, provided that the operating limits of the tube do not get exceeded.

Voltage Transformer : An instrument transformer which is used for the transformation of voltage. Its operation and design have been similar to those of an ordinary power transformer. It may be single-phase or three-phase. It may be known alternatively as a potential transformer.

Voltaic Cell : A self-contained source of electric energy which consists of two electrodes immersed in an electrolyte.

Voltaic Current : Electric current which resulted from chemical action.

Voltameter : Instrument for measuring quantity of electricity; also known as a coulometer.

Voltmeter : An instrument which is used for measuring the voltage or potential difference between two points, one of which may be at earth potential. The value is normally shown on a scale. Most volt-meters actually measure a current passing through a fixed resistance, the instrument then comprising a millimmeter or microammeter with a high multiplier resistance in series, the combination being connected across the points to be measured.

Volume Resistivity : Refers to the resistance offered to a constant and uniformly distributed current by a conductor of unit length and unit cross-sectional area. Dimentionally, resistivity is having the unit Ω-m, Ω-cm, Ω-in, etc., depending on the unit length chosen to express the dimensions of the cube-shaped conductor.

V.T. : Voltage Transformer.

Vulcanised India Rubber : Rubber treated with sulphur or sulphur compounds to modify its physical properties. The product, which is used as a low-voltage-cable insulant, is largely unaffected by temperature changes and is elastic.

Wager Earth : A device which is used to eliminate false balance conditions in a.c. bridges caused by the effects of capacitance at higher audio frequencies between the

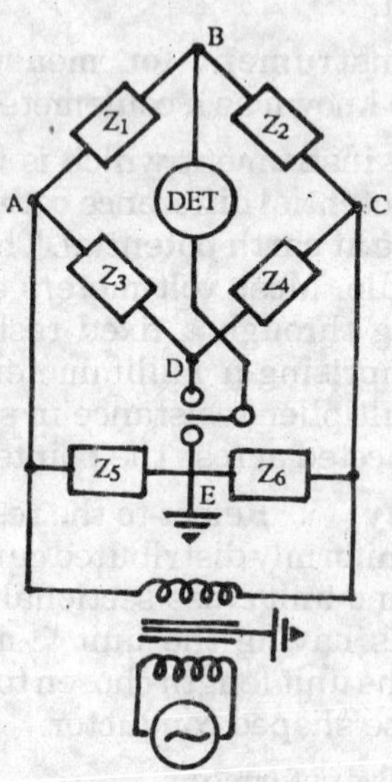

Fig. 1. Application of a Wager earth to an a.c. bridge. The ratio-arms Z_5 and Z_6,connected to earth, enable perfect balance to be obtained.

bridge arms and earth. If, as in Fig. 1, the bridge supply has not been earthed but the bridge is so arranged that points B and D are at earth potential, then the detector will also be at earth potential and no stray capacitance currents will flow in it. Further, all branch capacitance currents to earth will get stabilised. The Wagner earth method secures this condition by using two additional impedances Z_5 and Z_6, the junction E of which gets solidly earthed. Z_5 and Z_6 must be of the same type as either Z_3 and Z_4, or Z_1 and Z_2. If balance is achieved for both positions of the switch, then points B and D must have the same potential to earth as E, which is zero.

Walker Advancer : Refers to a phase advancer that resembles the Leblanc advancer with the addition of a stator winding in series with the armature winding. The electromotive force generated can be adjusted as regards its phase angle by altering the brush position relative to the stator winding.

Ward - Leonard Control : Refers to a form of motor control. The armature of the motor has been supplied from

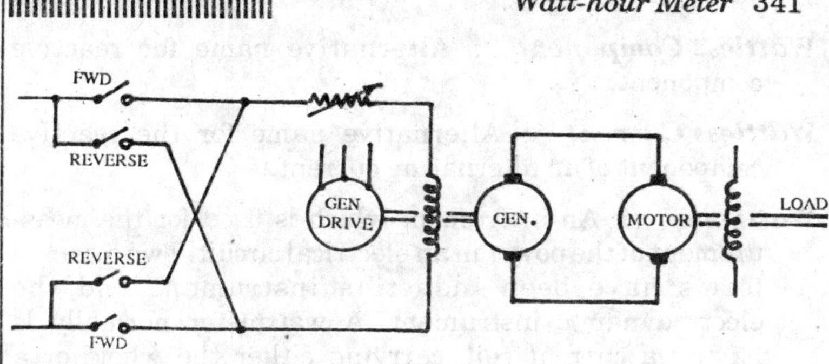

Fig. W. 2. Ward=Leonard drive.

its own generator,usually driven by an a.c. motor, as shown in Fig. 2. The field of the motor gets energised from another, and some time constant, source.

Ward- Leonard/Ilgner Control : See Ilgner system, Stubbs-Perry system.

Water Turbine : A prime-mover which used hydraulic power, *e.g.*, for driving electric generators. The design gets affected by the head of water available. An impulse turbine has been the Pelton wheel; a reaction type is the Francis turbine. A development of the reaction turbine is the propeller turbine or Kaplan turbine.

Watt : Refers to the SI unit of power (symbol: W). It is the rate of work, or energy transfer, at one joule per second. The joule has a mechanical definition, so that the watt has been primarily also a mechanical unit. It is electrically defined as the rate of transfer or conversion of energy when a current of one ampere flows between points having a difference of potential of one volt.

Watt-hour : Refers to unit of energy, being the work done by 1 watt acting for 1 hour. It equals 3600 joules.

Watt-hour Efficiency : Refers to the ratio between the amount of output energy in watt-hours drawn from a battery during a test discharge and the input energy in the same unit needed to charge the battery.

Watt-hour Meter : An integrating meter which is used for measuring energy. This is usually expressed in kilowatt-hours. It is alternatively called an integrating wattmeter or a recording wattmeter.

Wattless Component : Alternative name for reactive component.

Wattless Current : Alternative name for the reactive component of an alternating current.

Wattmeter : An instrument which is used for the measurement of the power in an electrical circuit. Two common forms have been induction instrument and the electrodynamic instrument. A wattmeter normally is having a current coil, carrying either the whole or a definite fraction of the load current, and a voltage coil carrying a current proportional of the voltage. In measuring the power in a single-phase or a d.c. circuit, the

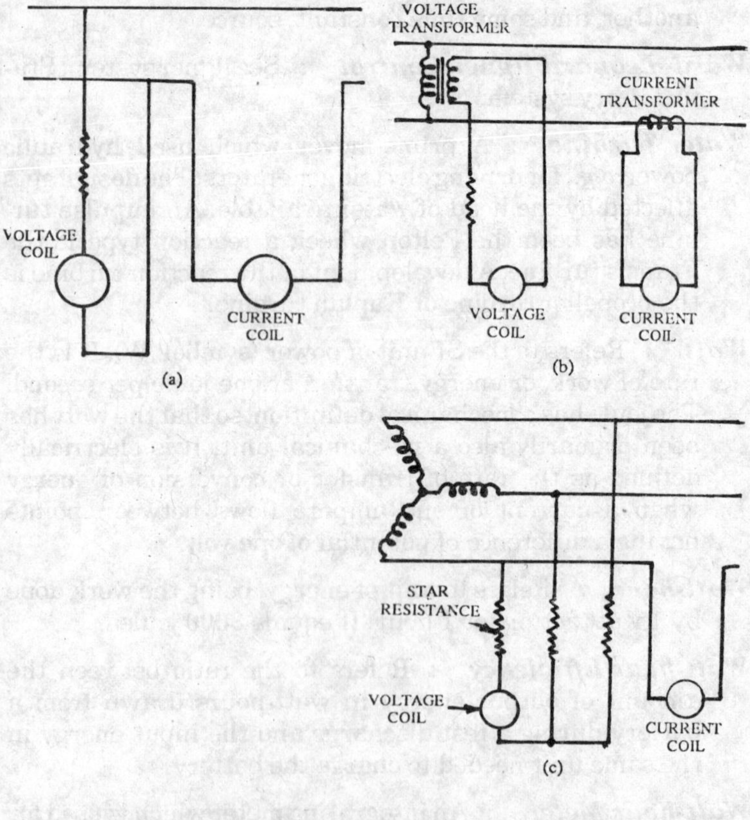

VOLTAGE COIL

CURRENT COIL

(a)

VOLTAGE TRANSFORMER

CURRENT TRANSFORMER

VOLTAGE COIL

CURRENT COIL

(b)

STAR RESISTANCE

VOLTAGE COIL

CURRENT COIL

(c)

Fig. 3. Three methods of connection for the voltage and current coils of a wattmeter : (a) simple connection, (b) high-voltage connection, (c) three-phase connection.

simplest form of connection is that given in Fig. 3(a). The current coil gets connected in one line and the voltage coil is connected between the lines. Where the current has been to high to pass directly through the coils of the instrument, and in the case of high-voltage circuits, current and voltage transformers are used as shown in (b).

Wauchope Starter : Refers to a star/delta starter for three-phase squirrel-cage induction motors which achieves a smooth acceleration approaching that of a slip-ring motor. Fig. 4 reflects the changes of connections in the Wauchope starter. The sequence of starting is (a) The motor gets connected in star and is permitted to accelerate to a stable speed in the normal manner. (b) The three resistors are then connected in parallel with the motor windings. This is simply a preparatory step and has been in operation for a fraction of a second only. (c) The star point of the windings has been now opened.

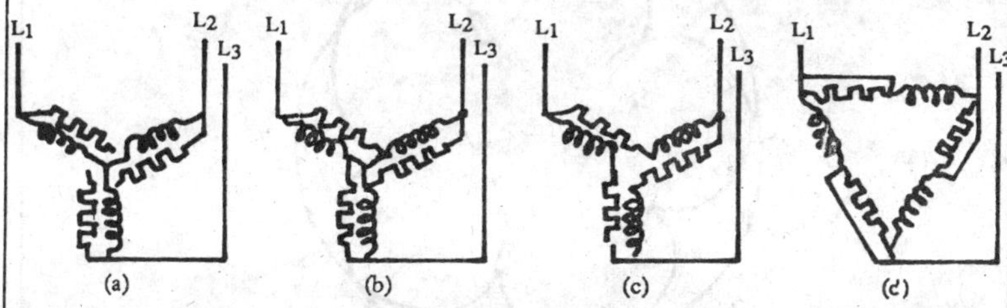

Fig. 4. Switching sequence of a Wauchope starter (Allen West & Co. Ltd.)

Wave Filter : Refers to a two-port network having a characteristic that enables it to discriminate between signals of various frequencies. It is a recurrent network (or artificial line) designed to transmit signals in a desired band (or bands) of frequency and attenuate those of other frequencies.

Wave-front : Refers to the rising portion of a voltage (or current) time characteristic of a surge voltage or current.

Wave-guide : Refers to a conductor of ratio waves of sufficiently high frequency constructed in the form of a metal or dielectric tube, or a single wire or a dielectric rod.

Wave Impedance : Refers to the characteristic impedance of a transmission line, or a measure of the impedance of a waveguide.

Wave-power Generator : Refers to a generator that employs sea waves as the source of power.

Wave-shape : Refers to the voltage or current/time characteristic of a surge. It may be defined by duration of the wave-front (T1) and the time to half-value of the wave-tail (T2), and hence is known as a T1/T2 wave.

Wave-tail : Refers to the falling portion of a voltage/time or current/time characteristic of an impulse voltage or current.

Wave Winding : Refers to an armature winding in which there have been only two parallel circuits through the armature, regardless of the number of poles. AS simple wave connection in five slots is shown in Fig. 5.

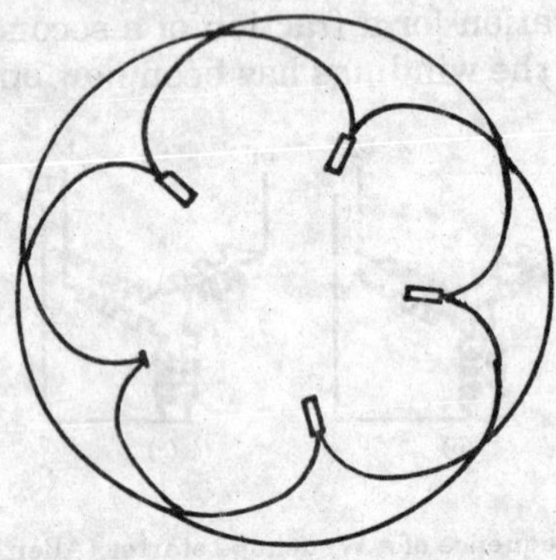

Fig. 5. Simple wave connection. The diagram shows a wave-wound armature with five slots and five bars.

Way : A space in a multiple duct unit to contain a cable.

Wayleaves : Refers to permission to carry transmission lines or telephone wires over private land or buildings; also the rent or charge for such permission.

Weber : The SI unit of magnetic flux (symbol: Wb). A weber per square meter is the unit of magnetic flux density. 1 Wb/m^2=10^4 gauss; 1 Wb=10^8 maxwells.

Weber Dynamometer : Early type of dynamometer, in which a small moving coil has been hung from a bifilar suspension inside a fixed coil.

Weld : Refers to a union between metal surfaces that get heated to a plastic or liquid condition.

Wenner Configuration : Refers to an arrangement of earth electrodes which consist of four electrodes in a straight line at equal intervals.

Werren Overlap Test : Refers to a modification of the Murray loop test for locating faults in cablest which is designed for use when the insulation of the return conductor has a low value, such as may often be found in a multicore cable where all cores are affected by the fault.

Weston Cell : A primary standard cell which is most commonly used. It has electrodes of mercury and cadmium amalgam in an electrolyte of cadmium sulphate with mercurous sulphate as depolariser (Figure 6). Its e.m.f. has been taken as 1.108 V at 20°C and the rate of change with temperature has been closely known.

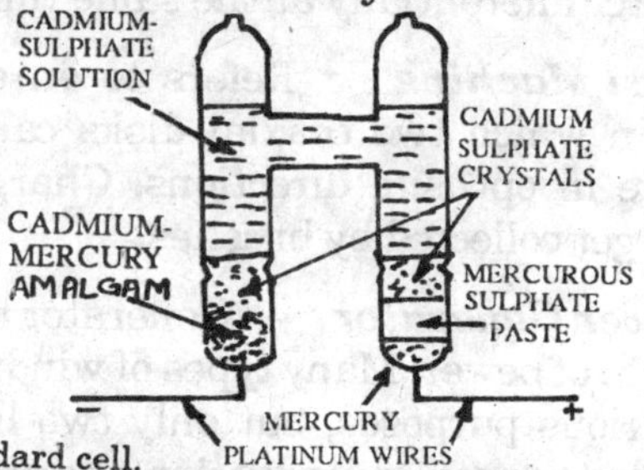

Fig. 6. Weston standard cell.

Wet Cell : Refers to an electrolytic cell, in which the electrolyte is a liquid. Wet primary cells are the Daniell cell and Leclanche cell.

Wetherill Separator : A magnetic separator which is used for removing a number of materials from a mixture. The material to be treated is passed on a conveyor belt between the poles of a magnet. The success of the machine depends on the shape of the magnet poles.

Wheatstone Bridge : Refers to a bridge circuit which consists of four resistance arms, a galvanometer and a battery or other d.c. source. If three of the resistances are known, the fourth could be determined in Figure 7, no current will flow in the galvanometer branch G if whence PR=QS is the balance

$$P_{i2}=S_{i2} \text{ and } Q_{i2}=R_{i2}$$

condition.

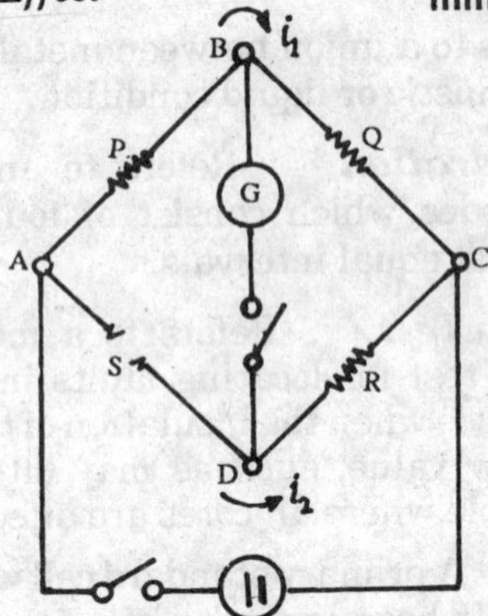

Fig. 7. Wheatstone bridge circuit. The galvanometer is undeflected if theratio of P to Q equals that of S to R

Wiedemann Effect : Refers to magnetostrictive effect in which a specimen twists when magnetised longitudinally and circumferentially at the same time.

Wimshurst Machine : Refers to an electrostatic generator, in which two coaxial disks carrying metal strips revolve in opposite directions. Charges induced on the strips get collected by brushes.

Wind-power Generator : Generator that use the wind as a source of power. Many types of windmill have been used for various purposes, but only two have been seriously regarded the power production, the propeller type and the Andreau generator.

Windage Loss : Refers to the power absorbed in a machine through incidental disturbance of the atmosphere by moving parts.

Winder : Refers to an electically-driven engine for hoisting a cage up the shaft of a mine or similar vertical shaft.

Winding : Refers to a system of insulated conductors forming the current-carrying element of a machine or transformer and designed to produce a magnetic field or be influenced by one. An electric machine operates because of the magnetic flux set up in its magnetic circuit by magnetomotive forces arising from currents flowing in groups of windings suitably disposed on the stator and rotor. The flux usually sets up an e.m.f. in the winding due to the conductors of the winding cutting the flux, or the turns of the winding being linked with a varying flux.

The interaction of the m.m.f.s. of the stator and rotor windings sets up a torque.

Winding Ends : Refers to the leads which are brought out from the two ends of a phase winding.

Wiping Gland : Refers to a watertight gland that gets achieved by the use of a wiped joint, or joint around which molten solder is wiped by hand with a cloth pad.

Wire Gauge : Refers to a means of specifying the size of circular-section wire or the thickness of sheet. Several sets up gauges have been in common use.

Withstand Test : A test which is applied to equipment designed to operate at high voltages. A specified alternating voltage has been applied to the test object for a specified time. This voltage must be withstood without flashover or puncture and may be carried out with the test object dry or subjected to artificial rain (wet test). it may be followed by a flashover test.

Work Function : Refers to potential difference traversed by an electron in performing sufficient work to permit it to leave the surface of a conductor. The work function gets varied with the material.

Wound-rotor Motor : Alternative term for a slip-ring motor.

Xenon : A range-gas element having atomic weight 131.3. It occurs in the atmosphere in one part in 170 million, by

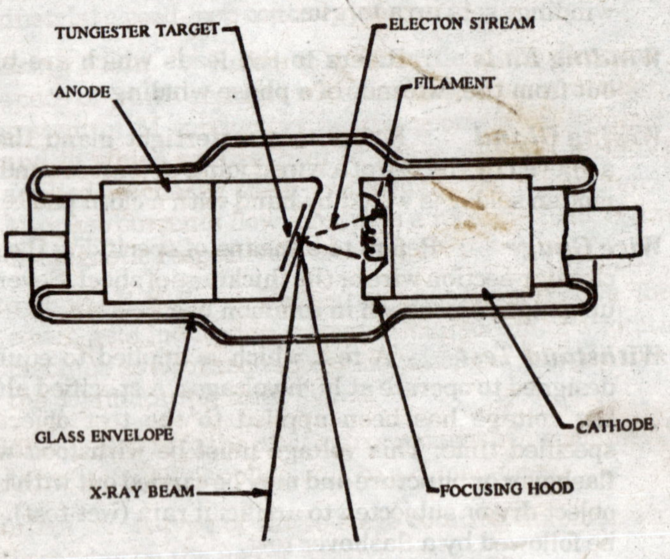

Fig. X. 1. X-ray tube.

volume. It finds use in some types of gas-filled electron tubes or discharge lamps.

Xerography : Refers to a dry photographic process which is based on electrostatics and triboelectricity, in direct contrast to other methods that need wet chemical processing for image development and fixation.

Xeroradiography : Refers to the production of an X-ray image by xerography.

Xfmr. : See transformer.

X-amplifier : Refers to the thermionic amplifier which is used for amplifying the voltage producing the horizontal deflection of the beam in a cathode-ray tube.

X-plates : Refers to a pair of flat parallel electrodes which are mounted vertically, side by side in a cathode-ray tube. A difference of potential applied between the two plates gives rise to the horizontal deflection of the beam.

X-rays : Electromagnetic radiation which formed part of the electromagnetic spectrum and occupying a band beginning at the short wavelength end of the ultra violet spectrum. The rays are produced in an X-ray tube. X-rays finds wide use in the fields of science, technology and medicine as a means of analysis, diagnostics and therapy. Those of low energy are loosely termed as soft, and those of high energy as hard.

X-ray Crystallography : Refers to the study of the sub-microscopic structure of materials by producing of interference patterns which are resulted from the diffraction of X-rays by atomic planes (X-ray diffraction).

X-ray Spectrometer : An instrument which is used for determining the wave-lengths of X-rays and the relative intensities of various wave-lengths in an X-ray spectrum.

X-ray Tube : Refers to a two-electrode vacuous device (internal pressure less than 10^{-6} mmHg). It is fitted with a pure tungsten filament, from which electrons get thermionically emitted, as the main component of the cathode, and an anode of suitable design to serve as a target for the stream of electrons which are passing from cathode to anode when a high voltage of appropriate polarity has been applied between the two electrodes (Fig. 1). The high-velocity electrons impinging on the metal target cause X-rays to get emitted.

X-unit : An obsolete unit for wavelengths of electromagnetic waves. It is approximately 10^{-7} μm.

Y-amplifier : Refers to thermionic amplifier which is used for amplifying the voltage producing vertical deflection of the beam in a cathode-ray tube.

Y-connection : This term is used for a three-phase star connection.

Yoke : Refers to the part of the core of a transformer or the ferromagnetic material of an electromagnet which does not get surrounded by the windings. It acts to join the magnetic poles or to complete the magnetic circuit.

Yoke Permeameter : An instrument which is used for measurement of the magnetic characteristic of a ferromagnetic material. A bar of the material is kept in a made-up magnetising coil, and the magnetic circuit gets completed by either one or two yokes of laminated, high-permeability material of large cross-section. B and H have been measured ballistically by using search coils close to the specimen.

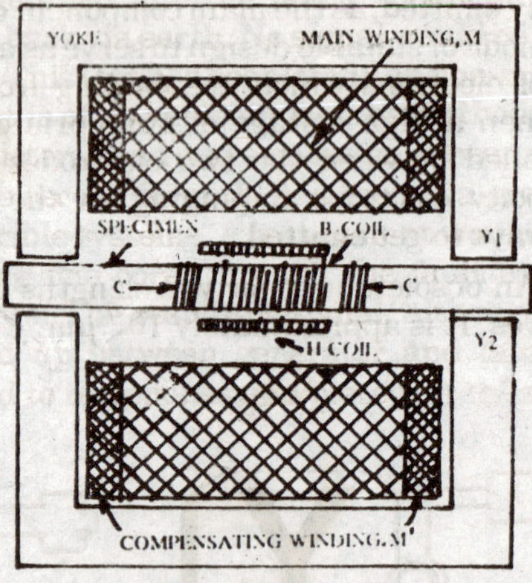

Fig. 1. A typical compensated yoke permeameter.

Yoke Suspension : The term used for the method of mounting a traction motor on an electric truck. It is more commonly known as bar suspension.

Y-plates : Refers to a pair of flat, parallel electrodes which are mounted horizontally, one above the other in a cathode-ray tube. A difference of potential applied between the two plates gives rise to vertical deflection of the beam.

Y-voltage : See star voltage.

Zener breakdown : Refers to a field-emission effect taking place when a high electric-field intensity exists across the depletion layer of a semiconductor. This field effectively gets reduced the forbidden energy gap, so that electrons could pass more easily from the valence band to the conduction band (see Figure 1).

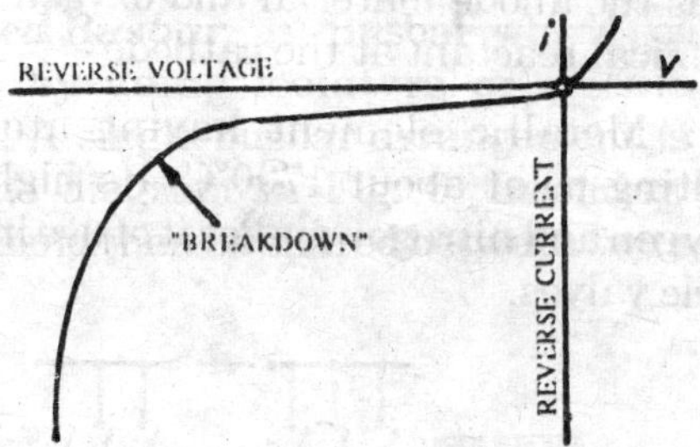

Fig. 1. Zener breakdown.

Zener Diode : The term has been now replaced by voltage regulator diode, and voltage reference diode.

Zener Effect : When a high reverse voltage gets applied to a p-n junction in a semiconductor there occurs a large reverse 'break-down' current, as shown in Figure 1. Breakdown limits the magnitude of the reverse voltage that could be applied to rectifiers and transistors.

Zero-pause : This term is used to refers to the momentary cessation of an alternation current when passing through a zero value between successive half-cycles. The zero-pause has been significant in the behaviour of circuit-breakers.

Zero Phase Sequence : Refers to a symmetrical phase sequence which corresponds to three equal co-phasal currents. It is possible to resolve any unbalanced system into three symmetrical systems, of positive phase sequence, negative phase sequence and zero phase sequence, which has been mutually independent.

Zero-type Dynamometer : Refers to the dynamometer in which the electric forces get balanced by mechanical forces to bring the indicating pointer to zero.

Zigzag Connection : Refers to a method of star connection in which each branch of the star is having contributions from two different phases.

Zigzag Leakage : Refers to the leakage flux which is produced in and near the gap between stator and rotor of an electric machine.

Zinc-air Battery : Refers to a dry primary battery which uses zinc as the anode material and oxygen as the other electrochemical reactant at the cathode.

Zirconium : Metallic element having atomic weight 91.22, melting point about 1700°C. Its high absorption rate for oxygen and nitrogen makes it of use in the making of electronic valves.

[illegible] for the reproduction in the receiver [illegible] sounds, like the voice of the speaker, [illegible] microphone of the same set. [illegible] by the use of hybrid coils.

[illegible] The SI unit of conductance and admittance [illegible]

[illegible] Dynamometer [illegible] Refers to a dynamometer [illegible] designed for the measurement of current [illegible] electromagnetic forces are balanced against [illegible] spring.

[illegible] of a lift car. This refers two [illegible] the car gets started from inside the car [illegible] from one switch, either on a [illegible]

[illegible]

[illegible] Refers to a [illegible] [illegible]

[illegible] commonly [illegible]

[illegible]